Uncertainty – The Pre-Resolved Strain

Sandeep Chavan

Published by Gyrus Vision, 2026.

UNCERTAINTY – THE PRE-RESOLVED STRAIN

First edition. February 24, 2026.

ISBN: 979-8233540349

Written by Sandeep Chavan.

Table of Contents

To those who seek clarity without demanding certainty.

Nothing is absolutely still.Nothing is absolutely certain.Yet everything persists.

Opening Note

Uncertainty has long been treated as a problem. In science, it appears as an error bar. In philosophy, it becomes doubt. In psychology, it manifests as anxiety. In spirituality, it is framed as mystery.

But what if uncertainty is none of these? What if it is not a defect in knowledge, not a limitation of tools, not a weakness of mind—but a structural feature of reality itself?

For years, I approached the world through engineering logic and the discipline of physics. I was trained to reduce ambiguity, tighten tolerances, and eliminate inefficiencies. Uncertainty was something to minimize, something to defeat. Yet the deeper I looked—into quantum behavior, black holes, time, consciousness, and even the everyday struggles of the human mind—I began noticing a pattern.

Nothing is absolutely still. Nothing is ever fully resolved. What we call stability is tension held in dynamic balance.

This book is not an attack on established science. It is not a rebellion against probability, nor a replacement for quantum mechanics. It is an attempt to reframe the very foundation beneath them. Instead of treating uncertainty as ignorance, I propose something simpler and more unsettling:

Uncertainty is pre-resolved strain within a continuous structure.

Reality is not made of isolated particles appearing and disappearing at random. It is a field under tension. What we call "events" are moments when that tension crosses alignment thresholds. What we call "time" is the span required for that alignment. What we call "certainty" is merely the exhaustion of competing possibilities.

This perspective did not emerge in a laboratory. It grew in classrooms, in discussions, in thought experiments, and in quiet observations of how systems—both physical and human—behave under load. It developed through a long journey into Universal Energy Dynamics and the concept of Alignment Threshold Time. But this book does not ask you to accept any framework blindly.

It only asks you to observe carefully.

When an object falls, when a wave collapses, when a decision is made, when fear rises—something is aligning. Something is resolving. Something was previously strained.

Uncertainty is not chaos. It is structure waiting for articulation.

If this book succeeds, it will not give you certainty. It will give you clarity about why certainty itself is limited. And perhaps that is enough.

— Sandeep

Preface

Uncertainty is often treated as a problem to be solved. In science, it appears as a limitation of measurement. In psychology, it takes the form of doubt. In society, it is seen as instability. In technology, it is framed as unpredictability.

The assumption remains consistent: with sufficient knowledge, refinement, and control, uncertainty will diminish. Yet across domains, uncertainty persists.

This book does not approach uncertainty as a statistical inconvenience or a philosophical abstraction. It examines uncertainty as structural continuity—a recurring condition observable in physical systems, cognitive processes, institutions, and technological architectures.

The central proposition is simple: uncertainty is not merely ignorance. It is not randomness. It is not error. It is the boundary condition of finite systems operating within dynamic equilibrium.

The framework presented here does not seek to replace existing scientific models or disciplinary methods. Instead, it identifies a shared geometry underlying them—finite alignment, temporal delay, residual strain, and scale-invariant incompletion.

The argument unfolds progressively, beginning with foundational structure, extending across domains, and returning to philosophical clarity.

This work is offered as architectural reframing—not as final doctrine, but as orientation.

How to Read This Book

This book is structured as a layered progression. The early chapters establish structural foundations—distortion, alignment, thresholds, and dynamic equilibrium. These concepts form the architectural base for all subsequent discussions.

The middle sections extend the framework across domains: quantum and physical systems, mind and cognition, social and institutional systems, and technological architectures. The purpose is not analogy, but continuity.

Later chapters synthesize the argument, repositioning uncertainty as fundamental rather than incidental.

The appendices provide precise definitions, cross-scale mapping, clarifications against misinterpretation, and operational orientation.

Readers may approach the book in two ways: sequentially, to follow the full structural arc; or selectively, to explore specific domains of interest while using the appendices as reference anchors.

This is not a book of prescriptions. It is a book of structure. It does not aim to eliminate uncertainty. It aims to make its geometry visible.

Prologue — The Illusion of Randomness

Why uncertainty feels like chaos. Why it is not.

We are raised to believe that randomness exists. A coin flips. A particle decays. A storm forms. A thought appears. A market crashes. A life changes direction without warning. We look at these events and say: *That was unpredictable.* And from unpredictability, we quietly conclude: *That was random.*

But unpredictability and randomness are not the same thing. Unpredictability is a limitation of perspective. Randomness is a claim about reality itself. The difference is subtle, yet profound.

From childhood, we associate uncertainty with instability. When something is uncertain, we feel uneasy. We cannot plan. We cannot control. We cannot forecast outcomes with confidence. The mind interprets this as danger. Psychologically, uncertainty feels like chaos because it denies us narrative comfort.

But chaos is not the same as disorder. A storm system looks chaotic from the ground; from above, it is geometry. Quantum behavior looks chaotic in measurement; in equation form, it is structure. Human emotions look chaotic in the moment; across time, they reveal patterns.

The feeling of randomness arises when our interpretive bandwidth is smaller than the structure we are observing. When alignment processes exceed our resolution capacity, we label them "random." This labeling is not wrong, but it is incomplete.

In physics, randomness entered the conversation most forcefully with quantum mechanics. The decay of a radioactive atom cannot be predicted precisely. The position of an electron cannot be determined without probability. Measurement outcomes appear fundamentally indeterminate. Many accepted this as proof that randomness is built into the universe.

But what if what appears random is not absence of structure—but presence of unresolved structure? What if uncertainty is not disorder, but strain?

A strained system does not behave smoothly. It oscillates. It competes. It delays. It holds multiple potential articulations simultaneously. To an observer with limited access, such a system appears indeterminate. Yet internally, it is not structureless. It is under tension. We confuse tension with chaos.

Consider something simple: a stretched elastic band held between two fingers. Before it snaps, there is uncertainty about when it will break. From our perspective, the exact moment is unpredictable. But the band is not random. It is under strain. Micro-fractures form. Energy redistributes. Internal geometry shifts. The delay before snapping is not magic—it is structural progression toward a threshold. The uncertainty exists because resolution has not yet occurred.

The same logic extends far beyond elastic bands. A decision not yet made. A wavefunction not yet measured. A black hole not yet fully described. A society under ideological stress. In each case, what appears chaotic is unresolved geometry.

Why, then, does uncertainty feel so deeply like disorder? Because the human mind demands articulation. We are uncomfortable with partial resolution. We prefer finality. The mind seeks closure because closure reduces cognitive load. Ambiguity consumes bandwidth. Competing possibilities require energy to hold.

So when we encounter unresolved states, the mind translates strain into anxiety. The emotional experience of uncertainty is not evidence of randomness. It is evidence of alignment still in progress.

This psychological response subtly influences our interpretation of physics itself. If something cannot be predicted precisely, we are tempted to declare it fundamentally random. It is cleaner. It is simpler. It ends the inquiry. But ending the inquiry does not necessarily resolve the structure.

There is another reason randomness seduces us: it preserves neutrality. If events are random, then no deeper structural explanation is required. If the universe contains irreducible indeterminacy, then unpredictability is an endpoint.

Yet nature consistently reveals something else: thresholds. Water does not boil randomly; it crosses temperature conditions. Materials do not fracture randomly; they accumulate stress. Stars do not collapse randomly; they exceed limits. Thresholds suggest geometry. Geometry suggests structure. Structure suggests process. Process implies delay. Delay implies unresolved strain. And unresolved strain is what we call uncertainty.

None of this denies probability. Probability remains a powerful descriptive tool. It models distributions, tendencies, and outcomes. But description is not ontology. Probability tells us how likely something is. It does not necessarily tell us what something is.

If uncertainty is pre-resolved strain within a continuous structure, then randomness becomes an appearance that emerges when articulation has not yet exhausted competing geometries.

In simpler terms: when a system has not crossed its alignment threshold, it appears uncertain. When it crosses, it appears certain. Between those states lies delay. And delay is not chaos. It is structure under negotiation.

This book begins with a simple proposition: the universe is not a collection of isolated, random events. It is a continuous field under dynamic tension. What we call "uncertainty" is not ignorance imposed upon us. It is the natural condition of finite systems resolving strain across time.

Nothing is absolutely still. Nothing is ever fully resolved. Stability is tension maintained gracefully within thresholds.

The illusion of randomness arises when we observe only fragments of a larger alignment process. Uncertainty feels like chaos because we stand inside the delay. But the delay itself has geometry. And geometry is never random.

Part I — The Structure of Distortion

"Nothing in reality is ever still. What we call stability is tension held within limits. What we call uncertainty is tension not yet resolved."

Overview

Before we speak of probability, quantum mechanics, psychology, or civilization, we must begin deeper. Uncertainty does not originate in mathematics; it begins in structure.

This first part lays the foundation. It introduces the simplest claim of this book: reality is not composed of isolated objects, but of continuous structure under strain. What we call "events" are moments of alignment. What we call "stability" is not absence of motion, but sustained dynamic equilibrium. What we call "uncertainty" is unresolved distortion within that equilibrium.

Here, appearance is separated from ontology. Continuity is examined. Distortion (*delta*) is defined not as external force, but as geometric strain within structure. We show why nothing is truly static, and identify the minimum conditions under which uncertainty must arise.

By the end of this part, randomness will no longer appear mysterious. It will appear structural.

Chapter 1 — Nothing Is Random

The Comfort of Calling Things Random

We call things random when we run out of explanation. A coin lands on heads five times in a row. A storm changes direction unexpectedly. A particle appears here instead of there. A person makes a decision we did not anticipate.

We pause. We shrug. We say, *"It's random."*

That word is not always a conclusion. Often, it is an exit. Randomness gives the mind relief. It closes the loop without demanding further structure. If something is random, it does not require deeper geometry, hidden interactions, or alignment processes unfolding beyond our sight.

But unpredictability is not the same as randomness. Unpredictability is about us—about the limits of our observation, our measurement, our bandwidth. Randomness is a claim about reality itself. When we label something random, we are no longer describing our ignorance; we are declaring that no deeper structure exists. That is a far stronger claim than we usually admit.

In daily life, the word *random* protects us from cognitive overload. The human mind prefers patterns. It wants causes. It wants narratives. When too many variables interact, when outcomes seem detached from visible inputs, we feel instability. The mind experiences that instability as discomfort. Calling something random stabilizes the narrative. It converts confusion into acceptance.

Psychologically, randomness is calming. It removes responsibility to investigate further. It reduces anxiety about prediction. It tells us: nothing could have been done; nothing deeper governs this.

But nature rarely behaves without structure.

A coin toss appears random because we cannot track the exact initial velocity, air resistance, torque, micro-imperfections, and landing surface dynamics. The unpredictability arises from complexity—not from absence of cause. Increase measurement precision enough, and the trajectory becomes calculable.

This example is familiar. It comforts us because it belongs to classical mechanics. The discomfort begins when we move to quantum behavior.

An electron does not have a precise position until measured. A radioactive atom decays at an unpredictable moment. Outcomes are described probabilistically, not deterministically. Here, randomness seems deeper. It appears woven into the fabric of physics itself. And so we accept it.

But acceptance is not explanation.

When something cannot be predicted with precision, we face two possibilities:

1. The structure is incomplete from our perspective.
2. The structure does not exist.

The second option is intellectually convenient. It ends inquiry. The first option demands patience. This book leans toward the first—not as rebellion against established physics, but as refusal to equate unpredictability with ontological chaos.

There is another subtle factor shaping our belief in randomness: scale. At small scales, interactions multiply beyond intuitive tracking. At large scales, systems exhibit sensitivity to initial conditions. Between micro-complexity and macro-instability, outcomes diverge rapidly from expectation. To a finite observer, such divergence feels indistinguishable from disorder.

But divergence does not imply absence of structure. A turbulent river is not random. A fractal coastline is not random. A market crash is not random. A human emotional reaction is not random.

In each case, strain accumulates. Competing influences interact. Thresholds are crossed. Articulation occurs. We tend to see only the moment of articulation. We miss the long preparation beneath it. The habit of calling things random emerges from observing outcomes without perceiving underlying tension.

There is also an emotional dimension. Uncertainty unsettles identity. When events cannot be controlled or predicted, the mind feels vulnerable. Randomness shifts responsibility outward. If outcomes are random, then we are not misaligned—we are merely unlucky. If reality itself contains irreducible chaos, then instability is inevitable and impersonal.

Randomness protects the ego. But protection is not the same as truth.

If reality is continuous—if nothing exists in isolation—then every event must participate in structure. Participation does not mean predictability. It means connection. Even if we cannot compute every interaction, continuity implies relational coherence.

A truly random event would require disconnection from all surrounding structure. It would require isolation from continuity itself. That is a very strong ontological claim.

Before accepting randomness as fundamental, we must ask: does nature ever truly disconnect anything? Or do we simply observe systems whose internal strain has not yet resolved within our perceptual window?

This distinction changes everything. If randomness is real, then uncertainty is ultimate. If randomness is illusion, then uncertainty is structural. The difference between these two positions will shape everything that follows in this book.

So we begin with a simple but destabilizing proposal: perhaps nothing is random. Perhaps what appears random is tension not yet articulated.

And if that is true, then uncertainty is not chaos. It is structure waiting to align.

1.1 — SVE as Continuity

If we are going to question randomness, we must begin beneath events—before particles, before probability, before outcomes. We must begin with continuity.

Modern physics often describes reality as a collection of entities: particles interacting in space, fields fluctuating in vacuum, discrete quanta exchanging energy. The language is convenient and computationally powerful, but language can quietly shape ontology. When we speak of "particles," we imagine separation. When we speak of "vacuum," we imagine emptiness. When we speak of "events," we imagine isolated occurrences.

Deeper investigation consistently reveals something else: continuity. The so-called vacuum is not empty. It is structured. It carries fluctuations, constraints, potentials. What appears as empty space is an active background. Nothing is ever truly detached from it.

In this framework, we refer to that baseline as **Structured Vacuum Energy (SVE)**—not as a mystical construct, but as disciplined acknowledgment that the background of reality is not void. It is structured continuity.

Continuity means there are no absolute gaps. Every local configuration participates in the larger structure. Every apparent "object" is a stabilized region within a continuous field. Nothing floats independently. Nothing emerges from nothing. Nothing vanishes into nothing.

A ripple in water does not exist apart from water. A vortex does not exist apart from air. A magnetic domain does not exist apart from its lattice. Likewise, what we call particles are localized excitations—configurations of a continuous substrate. The configuration may appear discrete; the underlying structure is not.

This has direct consequences. If reality is continuous, then isolation is an illusion of scale. When we treat an electron as an independent entity, we simplify for calculation. When we draw a boundary around a system, we define a model. But in ontology, boundaries are approximations. The structure does not stop at the edge of our equation.

Continuity implies that every distortion propagates. Not necessarily in dramatic ways, not always in ways we can measure, but structurally, no change is ever disconnected. A local distortion alters surrounding geometry. That altered geometry influences further alignments. Even stability is participation in continuous negotiation.

Now consider what randomness would require. For an event to be truly random in an ontological sense, it must occur without relation to surrounding structure. It must be uncaused not just by our ignorance, but by absence of continuity itself. But if continuity is fundamental, then absolute isolation is impossible.

This does not mean we can predict everything. Continuity does not guarantee computational tractability. It guarantees connection. Connection is different from predictability. A chaotic system is continuous, yet unpredictable in practice. A turbulent flow is continuous, yet sensitive to micro-variations. A nonlinear system can amplify microscopic differences into macroscopic divergence.

Unpredictability emerges from complexity interacting within continuity. It does not require disconnection.

When we observe a radioactive atom decaying at an unpredictable moment, we say the decay is random. But what if the atom is not isolated? What if it participates in background structure? What if decay is not spontaneous emergence from nothing, but threshold crossing within continuous strain geometry?

We may not yet possess full structural description. But absence of description is not proof of absence of structure. Continuity demands humility. It tells us that what appears independent may be locally stabilized configuration within larger relational geometry. It tells us that what appears sudden may be culmination of invisible interactions. It tells us that what appears random may be unresolved participation in structure beyond our observational bandwidth.

Structured Vacuum Energy is not an additional layer imposed on physics. It is recognition that the "empty" background is active and relational. It is the stage upon which all configurations exist—and not a passive stage, but an interacting one.

In such a reality, events are not isolated dots. They are reconfigurations of a continuous field. And if events are reconfigurations, then randomness cannot mean absence of relation. At most, it can mean incomplete access.

Continuity does not eliminate uncertainty. It reframes it. Uncertainty becomes not a break in structure, but a delay in articulation within structure.

Before we define uncertainty, before we introduce thresholds or alignment spans, we must anchor this idea firmly: nothing exists alone, nothing acts alone, nothing resolves alone. If reality is continuous, then randomness as ontological disconnection cannot exist.

What remains to be explained is not why events occur randomly, but why continuity under strain produces outcomes that appear unpredictable. For that, we must understand distortion itself. And that brings us to delta.

1.2 — Delta as Geometric Strain

If continuity is fundamental, then disturbance must be redefined.

In ordinary language, we speak of forces pushing objects, particles colliding, energy being transferred from one location to another. These descriptions are useful, but they subtly imply separateness—something acting upon something else, something moving through something else.

But in a continuous structure, there is no "outside" from which a force can arrive. There is only configuration and reconfiguration. Within continuity, what we call a disturbance is not an external intrusion. It is a distortion of geometry—a deviation from local equilibrium, a region where balance is perturbed and tension emerges.

In this framework, we call that distortion **delta**.

Delta is not a particle. Delta is not a force. Delta is not an entity traveling independently. Delta is geometric strain within continuity.

Imagine stretching a fabric. The stretch is not a separate object inserted into the cloth. It is a condition of the cloth itself. The strain is relational—it exists because the geometry has deviated from balance. Similarly, in a continuous energetic substrate, local distortions create tension fields. These distortions interact, compete, reinforce, or neutralize one another. But none of them are disconnected from the background. They are configurations of it.

This distinction matters. When we treat disturbances as objects, we search for where they begin and where they end. When we treat them as strain, we ask different questions: What geometry is distorted? What constraints prevent immediate realignment? What competing distortions are present? What threshold must be crossed for reconfiguration?

Strain implies persistence. A distortion does not vanish simply because we cannot see it. It persists as long as geometry remains misaligned. The system holds tension. It negotiates internally. It redistributes alignment possibilities. This is where uncertainty begins to take shape—not as ignorance, but as unresolved strain.

Consider a stretched elastic band again. Before it snaps, there is uncertainty about the precise moment of breakage. But the band is not waiting randomly. It is accumulating micro-strains. Fibers are redistributing load. Internal geometry is shifting. Competing microscopic distortions are interacting. The uncertainty lies not in absence of structure, but in the complexity of strain resolution.

Delta behaves similarly at all scales. In atomic systems, distortions appear as excitation states. In planetary motion, distortions appear as gravitational tension. In materials, distortions appear as stress concentrations. In human psychology, distortions appear as unresolved cognitive tension. In each case, what appears as event emerges from strain crossing a threshold.

But before threshold crossing, multiple alignment pathways coexist. Geometry is not yet committed. Competing deltas hold the structure in temporary negotiation. From inside that negotiation, outcome appears indeterminate. From outside, we call it random.

The deeper insight is this: delta does not "drive" reality. Nothing drives reality from outside. Delta simply expresses deviation within continuity. It is the condition under which alignment becomes necessary.

When a distortion forms, the surrounding structure must respond. That response is not instantaneous. Realignment requires propagation across geometry. Constraints slow articulation. Competing strains interfere. This delay is not accident. It is structural.

Until resolution occurs, the system contains multiple viable articulation pathways. It holds potential outcomes not as abstract possibilities, but as geometric configurations competing for stability. This is why uncertainty is not mere ignorance. It is the presence of multiple strain geometries within finite alignment span.

The mind tends to interpret unresolved strain as instability. But strain is not disorder. It is directional tension. It is structure in motion toward reconfiguration. Randomness would imply absence of directional coherence. Strain implies orientation. Even when outcomes appear unpredictable, they emerge from geometry seeking balance.

Delta therefore reframes disturbance from object to condition. Instead of asking, *"What caused this particle to behave randomly?"* we ask, *"What strain configuration existed that had not yet articulated?"* Instead of asking, *"Why did this system suddenly change?"* we ask, *"What distortion accumulated until threshold was exceeded?"*

Instead of seeing indeterminacy as chaos, we see it as unresolved geometry. And once we see delta as strain, the next step becomes inevitable: if distortions exist within continuity, how does stability persist at all? If strain is everywhere, why does reality not collapse into permanent instability?

The answer lies in dynamic tension. And that is the next movement.

1.3 — Stability as Dynamic Tension

If continuity is fundamental and delta is geometric strain, then stability cannot mean absence of disturbance. It must mean something else.

In ordinary language, we equate stability with stillness. A stable building does not move. A stable orbit repeats predictably. A stable personality does not fluctuate wildly. Stability feels like calm, like rest, like completion. But this intuition is misleading.

True stillness does not exist in nature. What we call stable is a condition of balanced tension—a dynamic equilibrium maintained within constraints. It is strain distributed in such a way that no single distortion overwhelms the structure. Stability is not the removal of delta. It is the management of delta.

Consider an atom. It appears solid and persistent. Yet internally, it is a choreography of interactions—electromagnetic constraints, quantum states, oscillatory behavior. Nothing inside it is frozen. Stability arises because competing influences hold one another in structured balance. Remove one constraint, and configuration collapses. Increase strain beyond threshold, and alignment shifts. The atom is not static. It is tension harmonized.

The same is true of planetary motion. A planet in orbit is not resting. It is falling continuously toward a star while moving tangentially. The orbit persists because gravitational pull and inertial motion maintain relational equilibrium. Stability here is not absence of motion—it is sustained negotiation between competing geometries.

Even a standing structure, like a bridge or a column, is stable only because internal stresses are balanced. Materials under load distribute strain across their geometry. Engineers know this well: remove enough redundancy, concentrate enough stress, and "stable" systems fail suddenly. But that suddenness is an illusion. The strain was always present. It was being carried. Stability is carried tension.

This understanding dissolves the myth of stillness. When we observe something at rest—a table, a mountain, a calm lake—we are perceiving equilibrium across timescales beyond our sensory resolution. The molecular lattice vibrates. Thermal motion persists. Internal forces continuously counteract gravity. The appearance of rest is macro-scale averaging of micro-scale dynamism.

The same logic extends to living systems. A stable mind is not one without thoughts. It is one where cognitive and emotional strains are integrated without escalation. A stable society is not one without conflict. It is one where tensions are distributed and absorbed before they cross critical thresholds.

In all cases, stability is not absence of distortion. It is distortion constrained within safe geometry.

This introduces a critical shift: if stability is dynamic tension, then instability occurs not when distortion appears, but when distortion exceeds balancing capacity.

Uncertainty therefore emerges naturally within stability. Because stability is not final. It is maintained. It requires continuous adjustment. Competing deltas are always present. Equilibrium is negotiated moment by moment. Resolution at one scale may generate new distortions at another.

This explains persistence. Why does reality not collapse under constant strain? Because strain is rarely unilateral. Distortions interact. Some reinforce; others counteract. Many remain below articulation threshold. The system carries them without visible transformation.

Think of it as ripple-lock. In certain configurations, distortions interlock in ways that reinforce overall geometry. Atoms form stable structures. Molecules form lattices. Orbits form predictable paths. These are not strain-free states—they are strain-balanced states.

And balance does not imply finality. Even the most stable configurations are vulnerable to extreme conditions. Increase energy density. Alter constraints. Introduce competing distortions. When thresholds are crossed, alignment shifts. What was stable reconfigures. But that reconfiguration is not random. It is strain resolving under altered geometry.

Here lies the deeper consequence: if stability itself is dynamic tension, then uncertainty is not the opposite of stability. It is embedded within it.

A stable system contains unresolved micro-strains at all times. Most remain below threshold. Some accumulate. Some interact. When articulation finally occurs, we perceive event. But the event was not spontaneous. It was culmination.

This reframing has profound implications. We stop asking, *"Why did this happen suddenly?"* We start asking, *"What tension was being carried?"* We stop equating calm with absence of motion. We recognize calm as controlled oscillation. We stop treating certainty as final truth. We see certainty as temporary exhaustion of competing strains.

Nothing in reality is frozen. Everything persists by holding tension within limits. And once we accept that stability is dynamic tension, the final misconception collapses: the belief that stillness exists at all.

What appears still is simply resolution occurring faster than our perception can detect. That misinterpretation is our next inquiry.

1.4 — Why "Stillness" Is a Misinterpretation

We trust our senses too much. When something does not visibly move, we conclude it is still. When something does not change within our perceptual window, we call it stable. When nothing dramatic happens, we assume nothing is happening.

But perception is not reality. It is sampling. The human nervous system operates within limited bandwidth. We do not perceive atomic oscillations. We do not detect thermal vibrations directly. We cannot observe microscopic strain redistribution in real time. Our awareness integrates signals across time and space, compressing complexity into manageable experience.

Stillness is often a byproduct of compression.

Consider a spinning fan. At low speed, each blade is distinct. Increase velocity, and the blades blur. Increase further, and the fan appears like a solid disk. To perception, motion has vanished into apparent solidity. Nothing stopped moving; resolution rate simply exceeded perceptual resolution.

The same principle operates across scales. A wooden table appears solid and unmoving. At the atomic level, it is a vibrating lattice of interacting excitations. Molecules oscillate. Bonds stretch and relax. Energy redistributes constantly. Yet to us, it is stable and still. What changed was not the structure, but our scale of observation.

The illusion of stillness emerges when dynamic equilibrium occurs at frequencies beyond sensory detection. We mistake limited sampling for ontological absence of motion.

This misunderstanding extends far beyond material objects. A calm lake on a windless day appears motionless. Beneath the surface, molecular activity continues. Temperature gradients shift slowly. Microcurrents form and dissolve. The surface reflects the sky in quiet symmetry, and we interpret this as rest. But equilibrium is not absence of activity. It is activity organized within constraint.

Even in celestial systems, where motion is obvious, we misinterpret stability. A planet in orbit seems predictably stable across centuries. Yet gravitational interactions subtly shift trajectories. Tidal forces deform bodies. Radiation redistributes energy. Over long durations, orbits evolve. Stability is long-scale negotiation.

We tend to label systems as static when their alignment span exceeds our timeframe of concern. Stillness is therefore a relative description. It describes the relationship between resolution rate and observer bandwidth—not intrinsic absence of dynamics.

This insight dissolves a powerful assumption. If stillness does not exist in any absolute sense, then neither can absolute certainty. Certainty implies completed alignment—no further adjustment required. But in a continuously dynamic structure, alignment is ongoing. Even the most stable configuration carries residual strain, however minute. At larger scales, these strains may cancel out or remain below threshold. But they are present.

Uncertainty, then, is not the interruption of stability. It is its companion.

The reason we feel shocked by sudden change—fracture, collapse, breakthrough—is because we misperceived prior stillness.

We believed nothing was happening. In truth, alignment was evolving invisibly.

Take the example of material fatigue. A metal beam supports weight for years without visible change. Then one day it cracks. To the observer, failure appears sudden. But internally, micro-strains accumulated with each load cycle. Bonds weakened incrementally. The final fracture was not random—it was threshold articulation. The illusion lay in believing the system was static before failure.

This illusion is deeply embedded in human psychology as well. A person appears calm for months, then reacts explosively. Observers say, *"It came out of nowhere."* But emotional strain accumulates. Competing narratives build tension. Internal equilibrium shifts gradually until threshold is crossed. Stillness was never real. Delay masked dynamics.

Our misinterpretation of stillness leads directly to our misinterpretation of randomness. When we fail to see micro-dynamics, we call macro-events random. When we overlook accumulating strain, we call threshold crossing unpredictable. But unpredictability often reflects scale mismatch.

The universe does not operate in discrete jumps separated by true voids of activity. It operates in continuous modulation of strain. Resolution at one scale may be invisible at another. High-frequency negotiation may average out into apparent calm.

What appears stable is often rapid alignment. What appears sudden is often slow accumulation reaching articulation.

Stillness is therefore not a property of reality—it is a perceptual artifact. Recognizing this changes how we interpret uncertainty. If nothing is ever truly still, then uncertainty is not the sudden appearance of motion in an otherwise static world. It is the visible phase of ongoing dynamic tension.

The difference between stillness and motion, between certainty and uncertainty, becomes a matter of alignment rate relative to observation. Reality is never idle. It is continuously negotiating geometry. When that negotiation becomes visible, we call it change. When it remains beneath perception, we call it stability.

In truth, both are expressions of the same continuous process. And if stillness is misinterpretation, then randomness—often inferred from sudden change—must be reconsidered as well. What appears random may simply be dynamic tension crossing perceptual threshold.

Which leads us naturally to the structural consequence.

1.5 — The Structural Consequence

Now we gather the threads. If reality is continuous, if disturbance is geometric strain within that continuity, if stability is dynamic tension rather than absence of motion, and if stillness is merely a perceptual compression of rapid alignment—then something important follows.

Randomness cannot be fundamental. What appears random must be reinterpreted.

In a discontinuous universe—one composed of isolated fragments—randomness could, in principle, exist ontologically. An event could emerge without relational dependency. It could occur without geometric continuity linking it to surrounding structure. But in a continuous substrate, nothing exists in isolation. Every local configuration participates in a broader relational field. Every distortion modifies surrounding geometry. Every alignment alters constraint distribution. Continuity eliminates the possibility of absolute disconnection.

This does not mean every event is predictable. It means every event is connected. And connection implies structure.

When we observe an outcome without perceiving the chain of relational adjustments that preceded it, we call it random. But from within continuity, the event is not disconnected—it is the culmination of strain resolving across finite alignment span.

This is the critical shift. Randomness assumes absence of relation. Uncertainty assumes unresolved relation. The difference is profound. Uncertainty becomes structural rather than mysterious.

Consider again what we have established: delta is geometric strain. Strain persists until alignment occurs. Alignment requires propagation across structure. Propagation is not instantaneous. There is delay. Within that delay, multiple alignment pathways may coexist. Competing strain geometries interact. Thresholds are approached but not yet crossed. From an observer's standpoint, outcome remains indeterminate.

But indeterminacy here does not mean absence of direction. It means articulation has not yet exhausted alternatives. Certainty, then, is not the dominance of a single force. It is the exhaustion of competing strain configurations. Before exhaustion, uncertainty remains.

This reframes probability as descriptive, not ontological. Probability does not create randomness; it models the distribution of possible articulations under unresolved strain. In this light, uncertainty is not an error bar added to an otherwise precise universe. It is the natural condition of finite systems negotiating strain within continuity.

The structural consequence is unavoidable: uncertainty is not ignorance imposed upon us. It is the visible signature of alignment in progress. Where there is continuity, there must be distortion. Where there is distortion, there must be tension. Where there is tension, there must be delay. Where there is delay, there must be unresolved geometry. And unresolved geometry appears uncertain.

This also clarifies why certainty feels temporary. Once alignment occurs, strain redistributes. New distortions emerge at other scales. Resolution at one level generates new tension elsewhere. Stability is always local. Global finality never arrives. Nothing is ever fully resolved—not atoms, not stars, not minds, not civilizations. Every articulation becomes the baseline for further strain.

Randomness would imply absence of memory in structure. Continuity implies that every resolution leaves imprint. Events are not isolated flashes. They are reconfigurations within evolving geometry. The universe is not throwing dice in empty space. It is negotiating tension within structured continuity.

If this is true, then the real question is no longer whether randomness exists. The real question becomes: what exactly is uncertainty?

If it is not ignorance, and not ontological chaos, and not magical indeterminacy—then what structural condition gives rise to it?

We have dismantled the illusion of randomness. But we have not yet defined uncertainty. That is deliberate. Before defining it, we must understand the minimum structural condition that makes uncertainty inevitable—not as error, not as ignorance, but as consequence of continuity under strain within finite alignment span.

If nothing is random, then uncertainty must have architecture. And that architecture is where we go next.

Chapter 2 — What Uncertainty Really Is

The Word We Use Without Understanding

We use the word *uncertainty* almost casually. The weather is uncertain. The economy is uncertain. The diagnosis is uncertain. The experiment produced uncertain results.

In science, uncertainty is printed beside every measurement. In psychology, it is the source of anxiety. In philosophy, it is a limitation of knowledge. In physics, it becomes a principle. Yet rarely do we stop and ask: what exactly is uncertainty?

Is it ignorance? Is it lack of precision? Is it randomness? Is it incomplete information? Or is it something more structural?

Most of the time, when we say something is uncertain, we mean that we do not know the outcome in advance. The future is not fully determined within our awareness. Multiple possible articulations exist, and we cannot predict with certainty which one will manifest.

But notice what we are actually describing. We are not describing absence of structure. We are describing delay of resolution.

If I toss a coin, the result is uncertain before it lands. But that does not mean the coin lacks geometry. It does not mean forces disappear. It means articulation has not yet occurred.

If a doctor waits for test results, the diagnosis is uncertain. But the biological condition already exists. What is uncertain is not reality—it is its articulation within knowledge.

Even in quantum mechanics, where uncertainty is elevated to foundational status, the description often stops at probability.

Position and momentum cannot both be precisely known. Measurement outcomes are probabilistic. The system is said to be indeterminate. But again, what is being described? Not absence of interaction. Not absence of structure. But absence of articulation prior to measurement.

There is something deeply revealing here: uncertainty always appears before resolution. Before the snap. Before the decision. Before the measurement. Before the collapse. After articulation, certainty appears—at least locally and temporarily.

This suggests that uncertainty may not be a flaw in knowledge. It may be a condition of systems that have not yet crossed alignment threshold. The word itself hides this possibility. It frames the phenomenon psychologically: *we are unsure.* But what if uncertainty is not primarily about the observer? What if it is about the structure itself?

When a system carries strain that has not yet resolved, it contains multiple viable alignment pathways. Geometry is not yet committed. The system is negotiating. From inside that negotiation, outcome appears indeterminate. But indeterminacy is not the same as chaos.

We often confuse the two because the mind prefers completed states. The mind is uncomfortable with partial resolution. It wants closure. It wants clarity. It wants outcome. When outcome is delayed, we call the state uncertain. Yet delay is not absence of structure. In fact, delay may be evidence of structure.

If alignment were instantaneous, uncertainty would not exist. Distortion would resolve immediately. No competing pathways would persist. No temporary multiplicity would arise. The very existence of uncertainty implies finite alignment span.

Something is happening between distortion and articulation. Something is holding tension. Something is preventing immediate collapse into a single configuration. That *something* is what this chapter seeks to define.

Chapter 1 removed the illusion of randomness. It showed that continuity eliminates ontological disconnection. But continuity alone does not explain why outcomes are not instantly determined. Now we must go further.

If nothing is random, then uncertainty cannot be chaos. If continuity holds, then uncertainty must have architecture.

This chapter begins that architecture. We will define uncertainty not as ignorance, not as mystical indeterminacy, but as a structural condition: **pre-resolved strain within finite alignment span.**

Before we accept or reject this idea, we must examine it carefully. Because if uncertainty is structural, then it is not something to eliminate. It is something inevitable.

2.1 — Pre-Resolved Delta

If uncertainty is not ignorance and not randomness, then what is it? To answer that, we must return to delta.

In the previous chapter, delta was defined as geometric strain within continuity—a distortion in local equilibrium. It is not a force arriving from outside. It is not an object traveling independently. It is a condition of imbalance within structured continuity.

Now we refine it further. Uncertainty exists when delta is present but not yet resolved. That is what we mean by *pre-resolved delta.*

When strain emerges in a continuous structure, alignment does not occur instantaneously. The distortion propagates. It interacts with other distortions. Competing geometries form. Threshold conditions are approached but not yet crossed. During this span—before articulation—the system carries tension. This is uncertainty. Not confusion in the observer. Not absence of cause. But unresolved strain within structure.

Consider a stretched elastic band once more. The moment after you begin pulling, the band is already under strain. Micro-level distortions redistribute tension across its geometry. You may not know when it will snap. But the snap is not random. It is prepared by accumulating micro-alignment shifts. Before fracture, the system exists in a pre-resolved state. That state is uncertain—not because it lacks structure, but because articulation has not yet occurred.

The same principle applies to decisions. Before committing to a choice, the mind holds competing narratives. Arguments for and against interact. Emotional and rational strains coexist. The decision has not yet articulated. There is internal tension. Multiple alignment pathways remain viable. This is psychological uncertainty. But again, it is not absence of structure. It is unresolved strain geometry within the cognitive field.

In quantum mechanics, a system prior to measurement is described as a superposition of possible states. Traditionally, this is interpreted probabilistically. But through the lens of pre-resolved delta, we can reinterpret it structurally: the system carries layered strain configurations. No single articulation has exhausted alternatives. Geometry is negotiating within finite alignment span. Measurement does not create structure. It exhausts competing strain geometries.

Before articulation, the system is not random. It is uncommitted. Pre-resolution means the geometry has not yet settled into a single stable configuration. It does not mean the system lacks internal structure. On the contrary, it means structure is active and unresolved.

This distinction is critical. Uncertainty is not the absence of a path. It is the coexistence of multiple viable paths within strain geometry.

Imagine water heating in a sealed container. As temperature rises, molecular motion intensifies. Pressure builds. Bonds strain. Before boiling occurs, the system exists in a pre-transition state. Phase change has not yet articulated, but internal conditions are preparing for it. That preparation phase is structurally uncertain.

At every scale, pre-resolved delta follows the same logic:

- Distortion appears.
- Competing configurations interact.
- Threshold is approached.
- Articulation has not yet occurred.

Uncertainty occupies that interval.

Notice something subtle here: pre-resolution is not static waiting. It is dynamic negotiation. Strain does not sit idle. It redistributes. It amplifies. It dampens. It competes. Geometry shifts continuously. But because no final alignment has yet been selected through threshold crossing, the outcome remains indeterminate from the observer's standpoint.

Indeterminacy therefore arises from ongoing dynamism—not from absence of causation.

This reframes the role of probability. Probability becomes a descriptive mapping of potential articulations within unresolved strain. It does not imply that structure is absent. It implies that articulation has not yet exhausted alternatives.

Pre-resolved delta also explains why uncertainty often feels tense. Because tension is literally present. In physical systems, strain accumulates energy. In biological systems, tension consumes metabolic resources. In psychological systems, unresolved cognitive strain generates anxiety. Uncertainty is experienced as discomfort precisely because it represents incomplete alignment.

The mind seeks resolution. The structure seeks balance. But balance requires crossing threshold. Until threshold is crossed, uncertainty persists.

Thus, uncertainty is not a defect in knowledge. It is not a glitch in reality. It is not mystical indeterminacy. It is the natural condition of systems carrying pre-resolved strain.

We can now say this clearly: uncertainty exists wherever delta has emerged but articulation has not yet occurred. This is the first structural component.

But rarely is there only one delta. Most systems carry multiple distortions interacting simultaneously. And it is their competition that deepens uncertainty further. That is where we turn next.

2.2 — Competing Strain Geometries

Pre-resolved delta alone does not fully explain uncertainty. Strain by itself creates tension. But uncertainty intensifies when strain is not singular—when multiple distortions coexist and interact within the same structural field.

Reality rarely hosts isolated disturbances. In a continuous substrate, distortions propagate. They overlap. They interfere. They reinforce or counteract. The geometry becomes layered. Multiple alignment pathways become simultaneously viable.

This is where uncertainty becomes structurally rich. When competing strain geometries coexist, no single configuration has yet exhausted the alternatives. The structure carries more than one potential articulation. Until one geometry stabilizes beyond threshold, the system remains indeterminate.

But indeterminate does not mean directionless. Each strain geometry represents a directional tendency within the field. Each attempts to stabilize according to its internal constraints and relational influences. They are not random possibilities. They are structured configurations competing for alignment dominance. Uncertainty persists when no single configuration achieves decisive stabilization.

Consider a flexible membrane pulled in two opposite directions. Two distortions exist simultaneously. The membrane deforms in response to both. The final configuration depends on relative strength, geometry, and constraint conditions. Before equilibrium is reached, the membrane does not "choose" arbitrarily. It negotiates. The negotiation phase is uncertain—not because the membrane lacks structure, but because the outcome depends on relational interaction between competing strains.

This same logic scales upward. In material science, a metal under complex loading may experience tensile and compressive stresses simultaneously. Micro-cracks may begin forming along multiple planes. Which crack propagates first depends on local geometry, imperfections, and strain concentration. Until propagation dominates, failure remains uncertain.

In human cognition, conflicting beliefs coexist. Emotional drives pull in opposite directions. Social pressures and personal values create competing narratives. Until one alignment pathway becomes dominant—until threshold commitment occurs—the person remains uncertain.

The presence of competition deepens pre-resolution. In quantum language, superposition can be reframed here. Rather than imagining a particle as existing in multiple mystical states, we can interpret the system as carrying layered strain geometries. Each configuration is structurally viable within the field. Measurement forces articulation by interacting with the system's geometry, collapsing competing strain possibilities into a single stabilized configuration. The system does not randomly jump. It resolves under interaction.

Uncertainty therefore increases when:

- Multiple deltas are present.
- Their magnitudes are comparable.
- Their constraints overlap.
- No geometry yet dominates decisively.

In such conditions, the structure remains dynamically unstable but temporarily balanced. This balance is not calm. It is tension distributed across alternatives.

Probability, in this view, becomes a map of geometric viability. It reflects how strongly each competing strain configuration aligns with surrounding constraints. A configuration with higher geometric coherence is more likely to stabilize. But likelihood is not randomness—it is weighted structural compatibility. The system does not roll dice. It weighs geometry.

Until one configuration exhausts competing alignments, uncertainty remains embedded in structure.

There is also an important insight here: uncertainty does not arise from infinite possibilities. It arises from finite but competing geometries. The system's configuration space is constrained. It is shaped by continuity, by boundary conditions, by relational structure. The competition is not arbitrary—it is structured.

Even when outcomes appear unpredictable, they emerge from bounded interaction among defined strain configurations. This explains why uncertainty is often temporary. As soon as one strain geometry exceeds stabilization threshold, alternatives collapse. Competing configurations lose viability. The field reorganizes. Certainty appears.

But certainty is not the absence of strain. It is the exhaustion of competition. After stabilization, new distortions will eventually arise. New competitions will form. Uncertainty will re-emerge.

Thus, uncertainty is cyclical within dynamic continuity. It arises whenever competing strain geometries coexist below articulation threshold.

This reframing also clarifies something subtle. If a system contained only a single distortion with no competing influence and infinite alignment speed, uncertainty would not exist. The system would resolve instantly. There would be no negotiation.

Uncertainty requires competition. It requires relational interaction among multiple strain geometries within finite alignment span.

We now see the structure more clearly:

- Pre-resolved delta creates tension.
- Competing strain geometries sustain indeterminacy.
- Resolution requires time.

Time here is not clock time in the abstract. It is alignment span required for geometry to propagate and stabilize.

This leads directly to the next structural component. If competing geometries exist, and alignment is not instantaneous, then uncertainty must persist for a measurable duration. That duration is not arbitrary. It is structural.

We call it **Alignment Threshold Time.** And it is the backbone of uncertainty.

2.3 — Finite Alignment Threshold Time (ATT)

If uncertainty were merely the presence of strain, it would disappear instantly. If competing geometries existed but alignment were instantaneous, the system would resolve the moment distortion appeared. There would be no delay, no negotiation, no indeterminacy.

But alignment is not instantaneous. This is the critical structural fact.

When distortion arises within continuity, its influence must propagate. Geometry must adjust. Constraints must respond. Competing strain configurations must interact. This process takes span—not because time is a mystical dimension enforcing delay, but because alignment requires relational reconfiguration across structure.

We call this span **Alignment Threshold Time (ATT).**

ATT is not clock time. It is not psychological time. It is not an abstract fourth dimension. It is the structural duration between distortion and articulation.

Whenever delta emerges, the system does not immediately snap into final configuration. Strain must redistribute. Constraints must recalibrate. Threshold conditions must be satisfied before a dominant geometry stabilizes. Until threshold is crossed, uncertainty persists. ATT therefore defines the window in which uncertainty exists.

Consider a taut rope pulled sharply at one end. The disturbance does not appear simultaneously at the other end. The deformation propagates. Internal forces communicate along the rope's structure. The final alignment emerges after transmission delay. In that span, the rope's configuration is evolving. Before full alignment, its geometry is neither fully old nor fully new. It is intermediate. It is negotiating.

Now scale this up. In materials, stress waves propagate at finite speeds. In electromagnetic systems, changes travel at light speed. In biological systems, signals propagate through neural pathways. In social systems, information spreads through networks. In cognitive systems, competing narratives integrate over processing cycles. Every system resolves within finite alignment span. Uncertainty lives inside that span.

If alignment were infinite in speed—if geometry reconfigured everywhere simultaneously—uncertainty would collapse into immediate certainty. There would be no delay between distortion and resolution. But finite propagation is fundamental to structured reality. And finite propagation guarantees temporary indeterminacy.

This reframes time itself. Time is often treated as an independent dimension in which events unfold. But here we treat alignment span as the operational core of temporal experience. What we call "time passing" is the observable manifestation of alignment processes resolving distortion.

ATT therefore is not an optional concept. It is the structural reason uncertainty can exist at all.

Uncertainty requires three conditions so far:

1. Continuity (no isolation).
2. Distortion (delta).
3. Competing strain geometries.

But these alone are insufficient. Without finite alignment span, competition would resolve instantly. The fourth condition is finite ATT.

This finite span ensures:

- Multiple strain configurations can coexist temporarily.
- Resolution is delayed.
- Articulation is threshold-based, not continuous smoothing.

Threshold matters. Alignment does not occur gradually into certainty. It often appears sudden because once threshold is crossed, competing geometries collapse rapidly. But that apparent suddenness hides the duration of pre-threshold negotiation. The articulation event—the snap, the decision, the measurement outcome—is simply the visible crossing of alignment threshold. ATT measures the invisible preparation.

This is why uncertainty is always temporary. It cannot be permanent in a finite system. Eventually, threshold is crossed. One configuration stabilizes. Alternatives dissolve. Certainty appears—locally and briefly. But certainty is not timeless. New distortions emerge. New competitions form. A new ATT span begins. Thus, uncertainty is cyclical and structural.

Now consider the quantum domain. When a quantum system is described as indeterminate prior to measurement, we can interpret that state as existing within its alignment span. Competing strain geometries persist because articulation has not yet been forced through interaction. Measurement interacts with the system, compressing ATT. It accelerates threshold crossing. It does not create randomness; it forces articulation.

The finite nature of ATT explains why indeterminacy is observed. If resolution were immediate, quantum uncertainty would not appear. But because alignment requires structural propagation—even at microscopic scales—temporary coexistence of configurations becomes inevitable.

This also clarifies something essential: ATT is not a metaphysical delay inserted artificially. It is the natural consequence of relational continuity. Nothing reconfigures everywhere at once. Even light has speed. Even force fields propagate. Even neural impulses require milliseconds. Finite alignment span is universal.

And finite alignment span guarantees that between distortion and articulation there must exist a region of unresolved geometry. That region is uncertainty.

Uncertainty is therefore not mystery. It is the structural effect of finite alignment threshold time within a continuous, strained system.

But there is still another constraint shaping uncertainty. Even if alignment span is finite, systems and observers resolve within limited bandwidth. And bandwidth determines how uncertainty is perceived, amplified, or suppressed.

That is our next layer.

2.4 — Limited Resolution Bandwidth

Even if distortion exists, even if competing strain geometries coexist, and even if alignment requires finite span (ATT), uncertainty would feel very different if resolution capacity were infinite. But it is not.

Every system resolves within limits. Every observer perceives within limits. These limits define resolution bandwidth.

Bandwidth, in this context, is the capacity to process, propagate, integrate, and articulate strain across structure. It is not merely sensory capacity. It is structural capacity.

We can separate this carefully into two layers:

1. **Structural Resolution Bandwidth** — the system's inherent ability to process and redistribute strain.
2. **Observational Bandwidth** — the observer's ability to detect and interpret alignment processes.

Both shape uncertainty.

Structural Bandwidth

Every system has finite integration capacity. A material can redistribute only so much stress per unit span before failure. A neural network can process only so much information per cycle. A computational system has finite clock rate and memory bandwidth. A biological organism has finite metabolic throughput.

Even physical interactions propagate at finite speed. No alignment spreads infinitely fast. The system's geometry imposes limits on how quickly strain can be resolved.

When strain complexity exceeds structural bandwidth, resolution slows. Competing geometries persist longer. ATT stretches. Uncertainty deepens. This is not randomness. It is capacity limitation.

A simple system with few distortions resolves quickly. A complex system with layered distortions resolves slowly. As complexity approaches bandwidth limits, indeterminacy becomes prolonged. In extreme cases—high strain density, competing configurations near symmetry—tiny variations can determine which geometry stabilizes. This does not imply absence of structure. It implies that resolution occurs near bandwidth saturation.

When bandwidth is stretched, articulation becomes sensitive. Sensitivity is not chaos. It is finite capacity under load.

Observational Bandwidth

Now consider the observer. Even if structural resolution is ongoing, perception integrates at finite sampling rates. Human vision averages over milliseconds. Neural cognition integrates across cycles. Instruments measure within limited precision.

When resolution processes occur faster than observational bandwidth, we perceive stability. When resolution processes occur near observational limits, we perceive uncertainty.

This explains why the classical world appears deterministic while the quantum world appears indeterminate. At macroscopic scales, strain dilution and aggregation reduce visible competition. Alignment occurs across vast numbers of micro-events. Resolution rate exceeds observational bandwidth. We see stable trajectories, solid objects, predictable outcomes.

At microscopic scales, structural bandwidth approaches resolution limits. Alignment span becomes comparable to observational interaction. Competing strain geometries remain visible to measurement. Indeterminacy appears fundamental.

The difference is not ontological randomness. It is bandwidth proximity. When observer bandwidth approaches structural bandwidth, uncertainty becomes visible. Probability emerges as a description of resolution under limited bandwidth.

Bandwidth and Perceived Reality

Consider a rapidly vibrating string. If vibration frequency exceeds perceptual sampling, it appears still. If frequency approaches sampling rate, flicker appears. If sampling improves, motion becomes clear. Uncertainty scales with bandwidth mismatch.

This also applies to psychological systems. A mind processing many competing narratives under limited cognitive bandwidth experiences anxiety. Decisions feel uncertain because integration capacity is saturated. The internal system has not yet resolved competing strain geometries within available processing cycles.

In organizations, when informational complexity exceeds communication bandwidth, uncertainty increases. Outcomes appear unpredictable. But the unpredictability arises from overload, not absence of structure.

Bandwidth limitation is universal. No system processes infinite strain instantly. No observer perceives infinite resolution simultaneously.

Uncertainty therefore intensifies when:

- Strain complexity increases.
- Competing geometries are balanced.
- Alignment span extends.
- Bandwidth approaches saturation.

In such conditions, indeterminacy is prolonged—not because reality is random, but because resolution capacity is finite.

The Structural Role of Bandwidth

Bandwidth completes the architecture of uncertainty. Continuity ensures connection. Delta creates strain. Competing geometries sustain negotiation. Finite ATT ensures delay. Limited bandwidth determines duration and visibility of indeterminacy.

Remove bandwidth limitation, and uncertainty collapses. Remove finite ATT, and uncertainty vanishes. Remove competition, and uncertainty reduces.

Uncertainty is not mystical. It is structural consequence of finite processing within continuous strain geometry.

This realization shifts the entire discussion. Quantum uncertainty is not chaos at the foundation of reality. Psychological uncertainty is not weakness of mind. Social uncertainty is not breakdown of order. They are all manifestations of finite bandwidth operating within strained continuity.

What remains now is to assemble these pieces into a clear minimum structural condition. If we remove any one component, uncertainty disappears. That synthesis is our next movement.

2.5 — The Minimum Structural Condition

We can now gather the architecture clearly. Uncertainty is not mystical. It is not a philosophical limitation. It is not a decorative feature added by quantum theory. It is the inevitable outcome of a specific structural configuration. Remove any one component of that configuration, and uncertainty disappears.

Let us restate the full condition precisely. Uncertainty requires:

1. **Continuity** — A connected substrate (no absolute isolation).
2. **Distortion (Delta)** — Local geometric strain within that continuity.

3. **Competing Strain Geometries** — Multiple viable alignment pathways.
4. **Finite Alignment Threshold Time (ATT)** — Resolution requires span.
5. **Limited Resolution Bandwidth** — Processing and propagation capacity is finite.

Only when all five coexist does uncertainty arise.

Remove Continuity → No Structural Relation If reality were truly fragmented—if events occurred in isolation without relational geometry—uncertainty would not even have meaning. There would be no structured pathway toward resolution. Randomness would be absolute, but structural uncertainty would be impossible because there would be no coherent field negotiating alignment. Continuity is the foundation.

Remove Distortion → No Tension If no strain exists, no alignment is required. A perfectly equilibrated system with zero disturbance has no reason to reconfigure. No delta, no negotiation, no articulation. No distortion, no uncertainty.

Remove Competition → Instant Resolution If only one strain geometry exists—no competing alignment pathways—then resolution is straightforward. The system reconfigures deterministically along the single viable path. Competition is what sustains indeterminacy. Without competition, uncertainty collapses.

Remove Finite ATT → No Delay If alignment were instantaneous—if distortion propagated infinitely fast across structure—then no pre-resolution span could exist. Articulation would occur at the moment of disturbance. No delay, no uncertainty. Finite alignment span is therefore essential.

Remove Bandwidth Limitation → Immediate Integration If structural and observational bandwidth were infinite, competing strain geometries would be integrated instantly. Every interaction would be fully processed. No temporary coexistence of alternatives would persist. Finite capacity ensures prolonged negotiation.

When all five conditions are present simultaneously, uncertainty becomes unavoidable. This is the minimum structural condition.

Notice something important: nowhere in this architecture do we require randomness. Uncertainty emerges naturally from finite resolution within strained continuity. It is not imposed from outside. It is not metaphysical indeterminacy. It is not ignorance disguised as ontology. It is structural.

This reframing has powerful implications. Quantum indeterminacy fits this model: microscopic systems operate near structural bandwidth limits, with competing strain geometries and finite alignment span. Indeterminacy appears fundamental because resolution processes are close to measurement scale.

Psychological uncertainty fits this model: competing cognitive geometries, limited neural bandwidth, finite processing span.

Social uncertainty fits this model: competing institutional strains, limited communication bandwidth, delayed systemic articulation.

Uncertainty is scale-invariant under this architecture. It is the signature of finite alignment.

This also explains why certainty feels temporary. Once articulation occurs, one strain geometry stabilizes. Alternatives collapse. But continuity remains. New distortions emerge. Competition reforms. ATT begins again. Bandwidth limits remain. Uncertainty returns.

Certainty is local exhaustion. Uncertainty is structural renewal.

This completes the definition. Uncertainty is **pre-resolved strain within finite alignment span under limited bandwidth inside continuous structure.**

Nothing mystical. Nothing random. Nothing chaotic. Only geometry negotiating equilibrium.

Now that we have defined what uncertainty really is, the next question becomes sharper: if this architecture holds, what are its implications for how we interpret quantum mechanics, collapse, and probability? Because if uncertainty is structural, then the foundations of indeterminacy must be revisited carefully.

That is where we move next.

From Structure to Implication

We have now defined uncertainty structurally. Not as ignorance. Not as error. Not as mystical randomness. But as the inevitable condition of continuous systems carrying distortion, negotiating competing geometries, resolving within finite alignment span, under limited bandwidth.

The architecture is complete. Continuity ensures relation. Delta introduces strain. Competition sustains indeterminacy. Finite ATT guarantees delay. Bandwidth limitation prolongs negotiation. Uncertainty is not an anomaly in this structure. It is its natural expression.

But definitions alone are not enough. If uncertainty is structural, then several long-held assumptions must be reexamined.

When quantum mechanics describes indeterminacy, is it describing randomness—or structural pre-resolution? When measurement collapses a wavefunction, is something magical occurring—or is alignment threshold being crossed? When probability is assigned to outcomes, is nature throwing dice—or are we mapping geometric viability within unresolved strain?

The implications are not minor. If uncertainty is structural, then collapse must be reframed. If uncertainty is structural, then probability is descriptive—not fundamental. If uncertainty is structural, then indeterminacy is not chaos—it is delay.

This does not overthrow existing mathematics. It reinterprets its ontology. The equations remain. The interpretation shifts.

We now move from defining uncertainty to testing it. What is the minimum condition under which uncertainty becomes unavoidable? How does collapse appear when strain geometry exhausts competition? Why does the quantum domain expose indeterminacy so clearly?

Having built the architecture, we must now examine its consequences in physics itself. The next chapter moves us from structural definition to structural inevitability. Because once uncertainty is understood properly, the foundations of indeterminacy begin to look very different.

Chapter 3 — The Minimum Condition

When Does Uncertainty Become Inevitable?

We have defined uncertainty. But definition alone does not give it weight. Now we must ask something sharper—something more structural: *When does uncertainty become unavoidable?*

Is uncertainty merely an artifact of complexity? Is it the result of insufficient measurement? Is it a temporary inconvenience in otherwise orderly systems? Or is it inevitable once certain structural conditions are met?

This is where the discussion becomes serious.

If uncertainty is optional, then perhaps with better instruments, better mathematics, or greater computational power, it could disappear. If it is accidental, then perhaps it belongs only to special domains—quantum mechanics, chaotic systems, psychological indecision.

But if uncertainty is structural, then it must arise automatically whenever specific conditions exist. It would not need to be inserted into theory. It would not depend on human ignorance. It would not rely on mystical indeterminacy. It would emerge the moment the structure satisfies a minimum configuration.

This is the pivot of this chapter. We are no longer asking what uncertainty feels like. We are asking what makes it inevitable.

Three Hypothetical Universes

To answer, imagine three hypothetical universes:

The first universe: everything is disconnected. Events occur without relational continuity. No structure binds them. No propagation links one configuration to another. In such a universe, uncertainty would not even have meaning. There would be no negotiation, no strain, no alignment. Only chaos. Not uncertainty—disconnection.

The second universe: everything is continuous, but alignment is instantaneous. Distortion appears and resolves everywhere simultaneously. No delay exists. No competition persists beyond a moment. In such a universe, uncertainty could not survive. The system would articulate immediately. Certainty would be constant.

The third universe: continuity exists. Distortion emerges. Multiple alignment pathways are viable. Resolution requires finite span. In that universe, uncertainty cannot be avoided. It appears automatically—not because something is broken, but because structure itself demands negotiation before articulation.

This is our universe.

We live inside finite alignment. We exist inside strain that does not resolve instantly. We observe systems where competing geometries coexist temporarily. We experience delay between distortion and outcome. And in that delay, uncertainty lives.

The Consequence

Uncertainty is therefore not a mysterious ingredient sprinkled into reality. It is not an anomaly in quantum theory. It is not a weakness of the human mind. It is the natural state of any structured system operating under finite conditions.

This realization shifts the entire conversation. Instead of asking *why* uncertainty exists, we must ask: *What are the minimum ingredients required for it to appear?*

If we can identify those ingredients clearly, then uncertainty stops being philosophical and becomes structural.

This chapter is not about interpretation. It is about inevitability. If certain conditions are present, uncertainty must emerge. And once we see those conditions clearly, we will understand why no finite system can ever escape it.

3.1 — Why Uncertainty Requires Structure

At first glance, uncertainty seems like the opposite of structure. Structure suggests order. Uncertainty suggests indeterminacy. Structure implies coherence. Uncertainty implies ambiguity.

But this intuition is misleading. Uncertainty does not arise in the absence of structure. It arises because structure exists under constraint.

Case One: No Structure

Imagine a universe with no continuity, no relational geometry, no propagation, no persistence. Events occur independently, without connection to surrounding configuration. In such a universe, uncertainty would not even be meaningful. There would be no strain to negotiate, no alignment to complete, no threshold to cross. There would only be disconnected occurrences. That is not uncertainty. That is chaos.

Uncertainty presupposes that something is trying to resolve. It presupposes tension within a coherent structure. It presupposes continuity that allows multiple alignment pathways to coexist. Without structure, uncertainty cannot even be defined.

Case Two: Perfect Structure with Instant Resolution

Now imagine a second extreme. Continuity exists. Distortion appears. But alignment is instantaneous. The moment strain emerges, geometry reconfigures everywhere simultaneously. No delay exists between distortion and articulation. In this universe, uncertainty again cannot exist. There would be no span in which competing geometries coexist, no pre-resolution phase, no indeterminacy window. Certainty would be constant. The system would always be in completed alignment.

Structured Tension

These two extremes teach us something important: uncertainty does not arise from lack of structure, nor from infinite structure. It arises from structured continuity operating under finite conditions.

Structure must exist—but it must not be infinitely fast. Continuity must hold—but resolution must require span. This is the only configuration in which uncertainty becomes meaningful.

In our framework, structure is not static arrangement. It is dynamic continuity under strain. Once distortion (delta) appears, geometry must respond. If response is instantaneous, uncertainty disappears. If response is impossible due to disconnection, uncertainty again disappears—replaced by chaos.

Uncertainty therefore requires a structured system capable of negotiation. It requires:

- A substrate that connects configurations.
- The ability to hold competing strain geometries.
- The necessity of threshold crossing before articulation.

Only a structured, continuous system with finite alignment span can host uncertainty.

Why This Matters

Many interpretations of uncertainty—especially in physics—quietly treat it as fundamental indeterminacy, almost as if reality itself is intrinsically ambiguous. But ambiguity without structure is meaningless.

If there were no relational geometry, we could not even speak of "possible outcomes." Possibility implies structure capable of supporting alternatives. Possibility presupposes geometry. This is subtle but decisive.

Uncertainty is not the absence of geometry. It is geometry that has not yet committed. Commitment requires structure. Negotiation requires structure. Threshold requires structure. Without structure, there is nothing to negotiate.

The Structural Necessity

We can now state something clearly: uncertainty is not a flaw in structured reality. It is a consequence of structured reality operating under finite conditions.

If structure exists and resolution is finite, uncertainty must arise. It is not an anomaly. It is not an exception. It is not limited to special domains. It is the natural condition of finite structured systems.

This realization prepares us for the next step. If uncertainty requires structure, then what specific structural ingredients are sufficient to make it inevitable? How many conditions are truly necessary?

As we will see, the answer is surprisingly minimal. Only two additional ingredients are required beyond continuity. And once they are present, uncertainty cannot be avoided.

3.2 — Competing Deltas + Finite ATT

We now reduce the architecture to its minimum. Continuity alone is not enough. Distortion alone is not enough. Uncertainty becomes inevitable only when two additional structural conditions coexist:

1. More than one viable strain geometry (*competing deltas*).
2. Finite Alignment Threshold Time (*ATT*).

If both are present, uncertainty must appear. If either is absent, uncertainty collapses.

Competing Deltas: The Necessity of More Than One Path

Imagine a system with a single distortion and no competing geometry. A stretched spring released in a vacuum with no constraints except its own elasticity. It contracts. There is only one viable alignment pathway. Resolution is straightforward. There is no uncertainty about the direction of resolution.

Now introduce competition. Add opposing forces. Add constraints. Add asymmetries. Add boundary conditions. Now the spring does not simply contract; it negotiates. Geometry must balance competing influences. Uncertainty appears the moment more than one viable alignment pathway exists.

This does not mean infinite possibilities. It simply means that more than one configuration can satisfy structural constraints within a certain span.

- In cognitive terms, uncertainty arises when a person sees

two viable decisions.

- In material terms, when stress can propagate along multiple fracture planes.
- In quantum terms, when multiple state configurations remain geometrically compatible prior to interaction.

Without competition, resolution is immediate. With competition, negotiation begins.

Finite ATT: The Necessity of Delay

Now imagine that competition exists—but alignment is instantaneous. Even with multiple viable strain geometries, if resolution propagated infinitely fast across structure, the dominant geometry would establish itself immediately. No pre-resolution span would exist. Competition would not sustain indeterminacy.

Uncertainty requires delay. Finite Alignment Threshold Time (ATT) ensures that competing geometries can coexist temporarily. It creates a span in which the system carries multiple viable configurations without immediate articulation. ATT is what prevents instant collapse into certainty.

The moment distortion arises, the system must propagate alignment across continuity. That propagation takes span. During that span, no single geometry has yet exhausted alternatives. That span is uncertainty.

The Minimal Pair

We can now express the minimum condition concisely:

If a continuous system contains more than one viable strain geometry and alignment requires finite span, then uncertainty is unavoidable.

Nothing else is required. We do not need randomness. We do not need mystical indeterminacy. We do not need incomplete knowledge. We need only competition + finite alignment.

Examples Across Scales

- **Material System:** A beam under load develops stress concentrations. Two potential fracture paths form. Micro-cracks begin along both. Until one path propagates decisively, the system is uncertain. The moment one crack exceeds threshold, articulation occurs. *Competition + finite propagation = uncertainty.*
- **Cognitive System:** A person considers two life choices. Both align partially with values. Emotional and rational strains compete. Neural processing requires time. Until one configuration stabilizes, uncertainty persists. *Competition + finite processing = uncertainty.*
- **Quantum System:** A microscopic configuration supports multiple strain geometries within the field. Interaction with a measurement device forces alignment. Before interaction, competing geometries coexist within finite span. *Competition + finite alignment span = uncertainty.*

The Logical Reduction

- Remove competition → no uncertainty.
- Remove finite ATT → no uncertainty.
- Keep both → uncertainty must exist.

It is not philosophical. It is structural.

Uncertainty is not something that occasionally appears in special domains. It is the natural outcome of competition under finite alignment.

This is the minimum condition. Everything else—probability distributions, wavefunction descriptions, statistical mechanics—is elaboration.

The core is simple: **multiple viable strain geometries negotiating within finite alignment span produce uncertainty.**

Once we see this clearly, the nature of resolution itself becomes sharper. Because uncertainty does not last forever. It ends. And how it ends—how threshold is crossed and articulation occurs—is the next essential piece.

We now turn to threshold and articulation.

3.3 — Threshold and Articulation

If uncertainty is sustained by competing strain geometries within finite alignment span, then we must ask: *How does it end?*

Uncertainty does not persist indefinitely in finite systems. At some point, articulation occurs. A fracture forms. A decision is made. A quantum measurement yields a definite outcome. A phase transition happens. A vote is cast. A bridge collapses. A thought crystallizes. Something crosses a boundary. That boundary is threshold.

Resolution is not gradual smoothing into clarity. It is often sudden, discrete, decisive. But the suddenness is deceptive—it hides preparation. Threshold crossing is the visible moment when one strain geometry exhausts the viability of alternatives.

The Dynamics of Threshold

During uncertainty, competing deltas coexist. Each represents a viable configuration within the field. Each carries strain energy. Each negotiates alignment under constraints. But viability is not permanent.

As strain redistributes and alignment propagates, one geometry may gain structural advantage. It becomes more compatible with boundary conditions. It reduces overall tension more effectively. It absorbs competing strain. When this advantage exceeds a critical margin, the system reorganizes rapidly. This is articulation.

Articulation is not creation from nothing. It is exhaustion of competition.

Illustrations Across Domains

- **Material fracture:** For a long time, micro-cracks form along multiple planes. None dominates. Uncertainty persists. Then, one crack propagates slightly faster. It concentrates strain more effectively. Competing cracks lose viability. Within a brief span, failure localizes. From outside, it appears sudden. Structurally, it is threshold articulation.
- **Cognitive decision:** Multiple narratives coexist. Arguments compete. Emotional weights shift. Then, one alignment crosses a coherence threshold. It integrates internal strain more effectively. Competing narratives collapse. Decision appears as a moment. Preparation preceded it.
- **Quantum collapse:** Measurement is often described as "collapse." But collapse is articulation under interaction. When the system interacts with a measuring apparatus, competing strain geometries can no longer remain viable. The interaction constrains geometry. One configuration stabilizes beyond threshold. Others dissipate. Collapse is

not mystical selection. It is structural exhaustion.

The Nature of Threshold

Threshold therefore defines the end of uncertainty. Uncertainty exists only below threshold. Once threshold is crossed, articulation occurs. Certainty appears—not as absolute truth, but as stabilized configuration under current constraints.

Threshold is not domination by force. It is not victory of one entity over others. It is structural inevitability once viability margins are exceeded.

Imagine a ball resting on a ridge between two valleys. Slight shifts determine which valley it falls into. Before tipping, uncertainty exists. After tipping, certainty appears. But gravity did not randomly choose. Geometry determined the outcome once threshold was crossed.

Threshold compresses delay into event. The articulation event is simply the visible manifestation of accumulated alignment reaching decisive margin. This is why uncertainty often feels tense—because the system is near threshold. And this is why articulation often feels irreversible—because once one strain geometry exhausts alternatives, reversing requires new distortion sufficient to reestablish competition.

Completion of the Dynamic

Threshold and articulation complete the dynamic. Competition sustains uncertainty. Finite ATT sustains delay. Threshold ends negotiation. Articulation stabilizes configuration.

Certainty is therefore not the triumph of power. It is the exhaustion of viable alternatives under constraint.

Understanding this reframes collapse across all domains. Nothing collapses randomly. Nothing resolves magically. Nothing articulates without preparation. Threshold crossing is structural inevitability once conditions align.

And once we understand articulation as exhaustion rather than domination, we must rethink certainty itself. Because certainty, as we commonly understand it, may not mean what we think.

That is where we turn next.

3.4 — Why Certainty Is Exhaustion, Not Dominance

We usually think of certainty as victory. One force wins. One idea triumphs. One state overpowers all others. Certainty feels like strength—like dominance, like clarity imposed on chaos.

But structurally, this interpretation is flawed. Certainty does not arise because one strain geometry becomes infinitely strong. It arises because competing strain geometries lose viability. Certainty is not domination. It is exhaustion.

The Dynamics of Exhaustion

During uncertainty, multiple viable configurations coexist. Each strain geometry has structural coherence. Each can potentially stabilize under given constraints. None has yet crossed threshold.

As alignment propagates through continuity, small asymmetries accumulate. One configuration begins integrating strain more efficiently. Competing geometries become increasingly strained under the evolving constraint landscape. Eventually, the alternatives can no longer sustain themselves. Their viability collapses.

At that moment, articulation occurs—not because the dominant configuration has become infinitely powerful, but because the others have become unsustainable.

This is a subtle but profound shift. Dominance implies power. Exhaustion implies structural inevitability.

Illustrations Across Domains

- **Material fracture:** When a beam breaks, we do not say the crack "won." The material's ability to redistribute strain was exhausted. Competing micro-cracks lost viability. One fracture path became the only remaining configuration capable of relieving strain. Certainty appears as failure of alternatives.
- **Cognitive decision:** When a decision is finally made, it is not because one thought annihilated all others by force. It is because competing narratives lost coherence under internal constraints. Their arguments became untenable. One configuration remained structurally sustainable. Decision is exhaustion of alternatives.
- **Quantum collapse:** Collapse does not mean a particle chooses arbitrarily. It means competing strain geometries cannot coexist under measurement interaction. The interaction constrains geometry so tightly that only one configuration remains viable. Collapse is structural elimination.
- **Social systems:** An ideology does not dominate because it is absolutely correct. It dominates when alternatives lose structural viability—economic, cultural, or institutional strain renders them unsustainable. Certainty is local stabilization after alternative exhaustion.

The Nature of Certainty

This reframing dissolves the myth of absolute certainty. Certainty is always contextual. It is always local. It exists only under current constraints. Change the constraints, introduce new distortion, and competition re-emerges.

What appears certain today may dissolve tomorrow—not because truth changed, but because structural conditions shifted. Certainty is temporary equilibrium. It is the visible moment when competition falls below threshold.

This also explains why certainty feels sudden. Exhaustion often happens gradually. Alternatives weaken incrementally. But once their viability drops below threshold, articulation compresses into a brief event. The visible shift appears abrupt, even though exhaustion unfolded over span. From outside, it feels like domination. From inside structure, it is inevitability.

The Cycle of Renewal

Understanding certainty as exhaustion also clarifies something deeper: there is no permanent certainty in finite systems. Continuity ensures new distortions arise. Finite ATT ensures delay. Bandwidth limitation ensures competition persists.

Once new strain geometries form, certainty dissolves back into uncertainty. This cycle is not a flaw. It is structural renewal. Certainty is never absolute truth imposed on reality. It is temporary stabilization under given conditions.

Recognizing this prevents a dangerous misunderstanding: the belief that once articulation occurs, the system is forever resolved. Nothing is ever fully resolved. Every articulation becomes the baseline for new strain. Certainty is exhaustion within context. And once we see this clearly, the inevitability of uncertainty becomes unavoidable.

Which brings us to the final consolidation. If structure contains continuity, distortion, competition, finite alignment span, and limited bandwidth—then uncertainty is not optional. It is guaranteed. That is the structural proof we now assemble.

3.5 — The Structural Proof

We can now assemble the argument without metaphor. Strip it down to logic.

If a system satisfies certain structural conditions, uncertainty must emerge. Not occasionally. Not accidentally. Not as philosophical ambiguity. But as direct consequence of geometry operating under constraint.

The Chain of Necessity

1. **Continuity exists.** No configuration is absolutely isolated. Every local distortion participates in relational structure.
2. **Distortion (delta) arises.** Strain appears within geometry. Equilibrium is locally perturbed.
3. **More than one viable strain geometry exists.** Multiple configurations can potentially stabilize under current constraints.
4. **Alignment requires finite span (ATT).** Resolution does not occur instantaneously. Strain must propagate and negotiate across structure.
5. **Resolution bandwidth is limited.** Integration capacity—both structural and observational—is finite.

If all five conditions are present, uncertainty is inevitable.

Testing by Removal

- **Remove competition:** If only one viable strain geometry exists, resolution is deterministic and immediate within finite span. No coexistence of alternatives. No indeterminacy window. Uncertainty disappears.
- **Remove finite ATT:** If alignment propagates infinitely fast, even competing geometries cannot coexist for any measurable span. The system articulates at the moment of disturbance. Uncertainty disappears.
- **Remove bandwidth limitation:** If strain integrates infinitely fast and observers perceive with infinite precision, competing geometries resolve immediately within alignment span. No indeterminacy is experienced. Uncertainty disappears.
- **Remove continuity:** If events are disconnected, there is no coherent strain negotiation at all. Only chaos—without structured uncertainty. Uncertainty disappears.

Only when all elements coexist does uncertainty emerge.

The Structural Necessity

This is not philosophical speculation. It is structural necessity.

Uncertainty is the natural state of:

- Competing alignment pathways
- Negotiating within finite propagation span
- Under limited integration capacity
- Inside continuous relational geometry

This proof does not depend on scale.

At **microscopic scales**, strain densities are high, bandwidth margins are tight, and ATT is comparable to measurement interaction. Uncertainty becomes visible as quantum indeterminacy.

At **macroscopic scales**, strain dilution and aggregation push alignment processes far beyond perceptual bandwidth. Certainty appears dominant. Yet even here, under high stress, competition re-emerges and uncertainty becomes visible again.

Uncertainty is scale-invariant. Its visibility depends on bandwidth and density.

The Core Realization

Uncertainty is not something added to reality. It is something that arises automatically when finite structured systems negotiate distortion.

The universe does not choose to be uncertain. It cannot avoid being uncertain. Because infinite resolution is impossible in finite structure.

This is the structural proof.

Uncertainty is not the opposite of order. It is the companion of finite order.

And once we accept that uncertainty is inevitable under minimal conditions, a deeper paradox emerges:

If uncertainty is structurally unavoidable, why does the classical world appear stable, predictable, and deterministic most of the time? Why does large-scale reality look certain while microscopic reality looks uncertain?

The answer lies not in abandoning uncertainty, but in understanding how scale transforms its appearance.

That is where we move next.

The Paradox of Apparent Certainty

We have completed the structural proof. Uncertainty is not optional. It is not philosophical decoration. It is not an artifact of ignorance.

It emerges automatically wherever:

- Continuity exists,
- Distortion arises,
- Competing geometries coexist,
- Alignment requires finite span,
- Resolution capacity is limited.

Under these conditions, uncertainty must appear. And these conditions are not rare. They define our universe. Which brings us to a paradox.

If uncertainty is structurally inevitable, why does most of the world appear certain?

Why do planets orbit predictably? Why do bridges stand for decades? Why do classical trajectories follow clean equations? Why does everyday experience feel stable and deterministic?

If uncertainty is fundamental, shouldn't chaos dominate?

Visibility vs. Existence

This is where misunderstanding begins. We confuse visibility of uncertainty with its existence.

Uncertainty does not vanish at large scales. It becomes diluted. Competing strain geometries aggregate. Alignment spans extend beyond perception. Bandwidth mismatches conceal negotiation.

At macroscopic scales, articulation rates exceed observational resolution. Stability appears dominant not because uncertainty disappears—but because it becomes statistically compressed.

Determinism, in the classical sense, is not absence of uncertainty. It is uncertainty averaged beyond visibility.

Structured Compression

This is a decisive shift. The classical world is not free from strain. It is saturated with distributed strain.

It is not free from competition. It is stabilized by large-scale averaging of competing geometries.

It is not instantaneous in alignment. It is slow enough that micro-uncertainty integrates into macro-regularity.

What appears deterministic is structured compression.

Preparing for Scale

This realization prepares us for the next movement. We have proven that uncertainty is inevitable. Now we must explain why it appears differently across scales.

Why does the microscopic domain expose indeterminacy openly, while the macroscopic domain hides it behind apparent certainty? Why does quantum theory emphasize uncertainty, while classical mechanics celebrates predictability?

The answer does not lie in two separate realities. It lies in density, scale, aggregation, and perceptual bandwidth.

Uncertainty is not confined to the quantum world. It is embedded in all structure.

The next chapter will show how scale transforms its appearance—and why what looks certain is simply uncertainty held in dilution.

We now move from inevitability to visibility.

Chapter 4 — Stability Is Not Absence

Why the World Does Not Collapse

We have argued that uncertainty is inevitable. Strain exists everywhere. Competing geometries coexist. Alignment requires span. Bandwidth is limited.

If all of this is true, then a natural question emerges:

- Why does the world not collapse?
- Why does reality not fragment under constant distortion?

Atoms persist for billions of years. Stars burn steadily for ages. Planets orbit in predictable paths. Biological organisms maintain coherence. Even identities endure across decades of internal conflict.

If uncertainty is built into structure, why does persistence dominate experience?

At first glance, this appears contradictory. Uncertainty suggests instability. Instability suggests fragility. Fragility suggests collapse. But collapse is not the rule. Persistence is.

The Misinterpretation of Stability

The mistake lies in equating stability with absence of distortion. We assume that a stable object is one without strain. That a stable orbit is one without tension. That a stable personality is one without internal conflict.

But earlier chapters dismantled this illusion. Nothing is ever free of strain. Nothing is ever absolutely still. Nothing is ever completely resolved.

So how does structure endure?

The answer cannot be that strain disappears. It cannot be that uncertainty vanishes. It cannot be that alignment becomes infinite in speed. Reality sustains under the very conditions that produce uncertainty.

This is the paradox we must now resolve. The persistence of structure does not contradict the inevitability of uncertainty. It depends on it.

Organized Tension

A system without strain would not adapt. A system without delay would shatter under disturbance. A system without negotiation would lack flexibility.

What appears stable is not frozen equilibrium. It is organized tension.

Consider something simple: a spinning top. It appears steady while rotating. If rotation stops, it falls. Motion sustains stability. Tension between angular momentum and gravity maintains posture. Stillness would mean collapse.

The same applies at deeper scales. An atom is not a static object. It is dynamic oscillation held in constraint geometry. Remove motion, and the configuration dissolves.

Stability, then, is not absence of activity. It is activity structured in such a way that strain redistributes instead of escalating.

Persistence Through Integration

This reframes our initial paradox. Reality does not collapse because strain is integrated. Uncertainty does not destroy structure because alignment capacity absorbs distortion before threshold is exceeded.

Persistence is not the absence of uncertainty. It is uncertainty organized.

This chapter explores that organization. We will examine how distortions interlock rather than destabilize, how local equilibria form within global dynamism, how delay itself becomes a stabilizing feature, and how stability is not the negation of strain, but its disciplined containment.

The world does not collapse because it is empty of tension. It endures because tension is structured. Understanding this dissolves the apparent contradiction between uncertainty and persistence.

And it prepares us to see stability not as stillness—but as dynamic equilibrium maintained across finite alignment span.

Now we examine how that equilibrium forms.

4.1 — Ripple Lock and Persistence

If strain is everywhere, and uncertainty is inevitable, then persistence must be explained in terms of strain—not in spite of it.

We begin with a simple but powerful idea: distortions do not always cancel. Sometimes, they interlock.

When multiple strain geometries constrain one another in a self-reinforcing pattern, the system stabilizes. Not because strain disappears, but because strain becomes organized. This is what we call **ripple lock.**

Standing Patterns

Imagine ripples in water moving toward one another. In certain conditions, they interfere destructively and flatten out. But in other conditions—under boundary constraints—they form standing waves. The surface appears patterned, even stable, but it is sustained by continuous oscillation. Nothing is static. The pattern persists because motion is structured.

Ripple lock operates similarly across scales.

- **Atomic systems:** An atom is not a static particle orbiting empty space. It is a stable configuration of constrained oscillation within structured continuity. Electromagnetic constraints and quantum boundary conditions create a geometry in which strain does not escalate—it cycles. The configuration persists because distortions interlock. Remove one constraint, and the structure dissolves. Alter geometry beyond threshold, and articulation occurs. Within its stable region, the atom is not strain-free. It is strain-balanced.
- **Orbital systems:** A planet orbiting a star is not resting in space. It is continuously falling toward the star while simultaneously moving tangentially. Gravity and inertia constrain one another. The resulting path is stable—not because forces disappear, but because they interlock. Motion sustains equilibrium. Remove tangential velocity, and collapse follows. Remove gravitational constraint, and dispersion follows. Stability arises from relational tension, not absence of influence.

Scaling Up

This principle scales upward. In ecosystems, predator and prey populations stabilize through dynamic feedback. In economies, supply and demand oscillate within regulatory constraints. In psychological systems, conflicting drives coexist under identity frameworks that prevent escalation.

Persistent systems are not devoid of distortion. They are configured such that distortion redistributes rather than amplifies. Ripple lock is therefore the structural explanation for persistence.

The Paradox Resolved

It answers the opening paradox: why does the world not collapse under constant strain? Because strain does not always accumulate unchecked. It often becomes self-limiting through relational geometry.

When distortions oppose each other in constrained balance, they prevent runaway escalation. They absorb one another's excess. They cycle rather than explode. Persistence emerges when competing geometries stabilize into mutual constraint.

Uncertainty and Stability

This is crucial. Uncertainty exists because competition and finite alignment span allow multiple strain geometries to coexist. Stability exists because those same competing geometries can enter lock conditions where none exceeds threshold.

Uncertainty and stability are not opposites. They are phases of the same dynamic. In uncertainty, competition remains unresolved. In stability, competition becomes organized. The difference lies in whether strain geometries amplify or interlock.

Resilience Through Ripple Lock

This also explains resilience. Systems that possess ripple-lock configurations can absorb new distortions without immediate collapse. New deltas enter the structure but are redistributed through existing constraints. Alignment span expands without threshold crossing.

Stable systems are not rigid. They are flexibly constrained. Rigidity breaks because it lacks redistribution pathways. Ripple-lock systems endure because they possess them.

Conditional Persistence

Now we see something subtle. Persistence does not require elimination of uncertainty. It requires that uncertainty remain below articulation threshold. As long as competing strain geometries remain balanced within constraint geometry, the system persists.

But ripple lock is always local. Under extreme strain, constraint geometry shifts. Redistribution fails. Competition intensifies. Threshold is crossed. Stability gives way to articulation.

Persistence is therefore conditional. It is sustained uncertainty under control.

And once we recognize ripple lock as the mechanism of persistence, we must ask: how do such stable regions form in the first place?

That leads us to local equilibrium formation.

4.2 — Local Equilibrium Formation

If ripple lock explains how strain can interlock into persistence, the next question is deeper: *How do such stable regions form at all?*

Why does distortion not spread endlessly through continuity? Why does strain not escalate uniformly across the entire structure?

The answer lies in **locality.**

Global equilibrium in a finite, dynamic universe is impossible. Distortion constantly arises somewhere. Alignment constantly shifts somewhere. If the entire structure had to reach perfect balance simultaneously, persistence would be unattainable.

Instead, stability emerges locally.

Local Equilibrium

Local equilibrium is not universal calm. It is a pocket of negotiated balance within broader dynamism.

Imagine a landscape of hills and valleys. A marble placed on that landscape will roll until it reaches a valley. It settles—not because gravity has ceased, but because the local geometry contains it. The valley does not eliminate gravity; it organizes it. The marble is stable locally, not globally.

This analogy captures something essential about structured reality.

The "valleys" in strain geometry are configurations where distortions redistribute in such a way that further escalation becomes energetically unfavorable. These are local minima in the strain landscape. They are not perfect solutions. They are stable under current constraints.

Atoms form because certain geometric configurations minimize relational strain. Molecules form because specific bonding arrangements reduce tension relative to alternatives. Crystals form because repeating patterns distribute distortion efficiently across lattice geometry.

In each case, stability is not the absence of force. It is the presence of locally optimal constraint.

Contextual Stability

Local equilibrium allows structure to exist without requiring total resolution of all strain everywhere. This is critical.

If every distortion required global realignment before stability could emerge, no structure could persist. Finite systems cannot resolve infinite relational complexity instantly. Instead, they form pockets of stability where strain is sufficiently integrated to avoid threshold crossing.

These pockets are conditional. Change boundary conditions, introduce new distortion, or alter constraint geometry, and the valley reshapes. What was once stable may become unstable. A new local equilibrium may form elsewhere.

This explains why stability is always contextual. There is no absolute equilibrium in a finite dynamic universe. There are only local equilibria sustained within specific constraint geometries.

Examples Across Scales

- **Planetary systems:** The solar system is not globally static. Gravitational interactions constantly perturb orbits. Yet stable orbital patterns emerge because the system settles into locally sustainable configurations. Small perturbations are absorbed; large perturbations may shift equilibrium.
- **Human systems:** Identities form as local equilibria of competing drives. Values, memories, and narratives create a configuration that distributes internal strain. Under new life events, this equilibrium may shift. But for long periods, it persists.

Local equilibrium is therefore not perfection. It is sufficient strain distribution under given conditions.

Dynamic Absorption

This reframes stability entirely. Stability does not mean all strain is resolved. It means strain is contained within bounds that prevent articulation.

Local equilibrium formation depends on geometry. When distortions can be redistributed across sufficient relational pathways, threshold crossing is avoided. When geometry lacks redistribution channels, strain concentrates and instability follows.

Thus, stable systems are those with geometry capable of absorbing distortion without amplifying it. And this absorption is dynamic. The equilibrium is maintained through continuous micro-adjustments. Even in apparently static systems, countless microscopic negotiations occur beneath perception.

Local equilibrium is not static balance—it is sustained recalibration.

Persistence as Controlled Uncertainty

Understanding this dissolves the fear that uncertainty implies inevitable collapse. Uncertainty exists within local equilibria as long as competing geometries remain below articulation threshold.

Persistence is simply the duration over which local equilibrium remains viable.

But local equilibrium does not freeze motion. It organizes it. Which leads to the next insight: reality sustains not by stopping movement, but by sustaining motion within constraint.

That is our next inquiry.

4.3 — How Reality Sustains Without Freezing

If stability is not the absence of strain, and local equilibrium is not global resolution, then another misconception must fall: the idea that persistence requires stillness.

We instinctively associate durability with rigidity. A solid object seems strong because it does not move. A calm person seems stable because they appear unshaken. A stable system seems reliable because it behaves predictably.

But in structured reality, freezing is fragility. A frozen system cannot redistribute new strain. It cannot adapt. It cannot absorb distortion. The next perturbation shatters it.

Reality sustains precisely because it does not freeze. It remains dynamically active at all times.

Motion as Stability

Consider again the atom. It persists not because nothing happens inside it, but because constant oscillatory motion redistributes strain. Quantum configurations fluctuate within allowed bounds. If the atom were truly static, it would collapse under disturbance. Motion sustains structure.

The same applies to biological organisms. The body remains alive because of continuous metabolic activity. Cells replace themselves. Proteins fold and unfold. Electrical signals pulse. If all motion ceased, structure would decay immediately. Stillness is not stability. It is termination.

This principle extends across scales. A spinning gyroscope remains upright because of motion. Remove angular momentum, and it falls. A flowing river maintains channel shape because of constant water movement. Stop the flow, and sediment accumulation reshapes the bed.

In each case, persistence arises from dynamic recalibration.

Continuous Redistribution

Reality sustains because distortions are continuously redistributed before they reach articulation threshold. This ongoing micro-resolution prevents catastrophic escalation.

Earlier, we defined uncertainty as pre-resolved strain within finite alignment span. Stability now appears as the condition where such strain remains below threshold through constant redistribution.

The key word is *constant.* Dynamic systems are always negotiating. Even a seemingly stable structure is processing micro-distortions continuously. Thermal fluctuations, environmental influences, internal interactions—all generate strain. But ripple-lock configurations and local equilibrium geometries absorb them.

This absorption is not passive. It requires motion. A rigid structure with no internal flexibility will accumulate strain until fracture. A flexible structure with dynamic pathways redistributes strain incrementally. Flexibility is structured dynamism.

Resilience Through Adaptation

This is why resilience and stability are not identical to rigidity. Rigid systems resist change but shatter under sufficient load. Dynamic systems yield slightly, absorb energy, and return to equilibrium. Sustaining requires adaptation.

From a structural perspective, persistence depends on the system's ability to continuously recalibrate within finite alignment span. New distortions emerge. Strain geometries shift. Local equilibrium adjusts. As long as these adjustments occur before threshold is exceeded, stability persists.

Delay as Strength

This is also why time matters. Sustained systems maintain enough alignment capacity over span to prevent rapid threshold crossing. They stretch uncertainty without allowing collapse.

A bridge flexes under wind. A building sways during seismic waves. A mind reinterprets new information. A society reforms policies in response to strain. Adaptation is dynamic stability. Freeze these systems, and collapse becomes inevitable.

Now we see the deeper truth: stability is not the negation of uncertainty. It is the management of uncertainty through continuous motion. Nothing in reality survives by becoming static. Everything that persists does so by remaining active within constraint.

Which brings us to a crucial insight: if delay is the span in which uncertainty exists, then delay is not weakness. Delay is structural strength.

We now examine how delay itself becomes measurable evidence of stability.

4.4 — Delay as Measurable Structure

We have already defined delay in structural terms. Delay is not passive waiting. It is alignment in progress.

Now we take a further step. Delay is not merely a consequence of finite alignment. It is an indicator of structural capacity.

In other words: the ability of a system to sustain delay without crossing threshold is a measurable sign of its stability.

Delay as Signal

When distortion enters a system, alignment must propagate. If the system lacks redistribution pathways, strain concentrates rapidly.

Threshold is crossed quickly. Collapse follows. Short delay under strain often signals fragility.

But when a system can absorb distortion, redistribute it across geometry, and extend alignment span without exceeding threshold, delay increases. Extended delay signals resilience.

Think of a bridge under load. A poorly designed structure will crack quickly when subjected to stress. Micro-fractures propagate rapidly. The time between disturbance and failure is short. A well-designed structure flexes. Internal stresses redistribute across beams and joints. The time between disturbance and failure extends dramatically. Under normal conditions, threshold is never reached. Delay here is measurable. It is not emptiness. It is structural negotiation.

Delay Across Systems

- **Materials:** Ductile materials deform plastically before fracture. They exhibit long delay between initial strain and catastrophic failure. Brittle materials shatter quickly. Their delay span is short. Delay reflects redistribution capacity.
- **Cognition:** A person with limited psychological bandwidth reacts impulsively. Distortion crosses threshold quickly. Articulation—often regrettable—occurs with minimal delay. A person with developed integration capacity holds competing narratives longer. They process tension without immediate collapse into reaction. The delay before articulation extends. That delay reflects internal structural strength.
- **Institutions and societies:** Systems that collapse immediately under pressure lack redistribution mechanisms. Systems that endure crises without catastrophic articulation possess alignment pathways

capable of extending delay.

Delay is measurable structure.

Physics of Delay

In physics, we see this in relaxation times, damping rates, decoherence times, phase transition durations. These are not arbitrary parameters. They measure how long a system can sustain pre-resolution span under distortion before articulation. Longer relaxation time under constraint often indicates more robust local equilibrium.

Delay is therefore not weakness. It is capacity.

ATT as Dual Measure

Earlier, we described Alignment Threshold Time (ATT) as the span required for articulation. Now we see its dual meaning. ATT is both:

- The window in which uncertainty exists.
- The measure of how much strain a system can absorb before crossing threshold.

Stable systems have meaningful ATT margins. Fragile systems have minimal ATT margins.

If alignment span is extremely short, articulation occurs immediately. The system cannot negotiate. It collapses under minor distortion. If alignment span is extended through redistribution geometry, the system sustains.

Time as Geometry

This reframes time itself in structural terms. Time is not an empty container in which events occur. It is the measurable span of alignment negotiation. Delay is geometry unfolding. And stability is the ability to maintain alignment negotiation without irreversible articulation.

Seen this way, persistence becomes quantifiable. The stronger the redistribution capacity, the greater the delay before threshold. The greater the delay, the more robust the local equilibrium.

This connects stability directly to uncertainty. Uncertainty is not eliminated in stable systems. It is stretched. The system carries competing geometries within extended alignment span without collapse.

But delay is not infinite. Under sufficient distortion, even robust systems cross threshold. Which leads us to the final synthesis: stability does not contradict uncertainty. It organizes it.

We now consolidate this paradox clearly.

4.5 — The Paradox Resolved

We can now resolve the tension that has been building since the beginning of this part.

Uncertainty is inevitable. Strain is everywhere. Competition persists. Alignment requires finite span. And yet, reality does not collapse. The resolution lies in a shift of perspective: **stability is not the absence of uncertainty. It is uncertainty organized within constraint.**

Earlier, uncertainty appeared threatening. If strain exists everywhere and competition is constant, then collapse should be frequent. But we now see that strain does not automatically escalate. It can interlock. It can redistribute. It can settle into local equilibria that sustain alignment below threshold.

Uncertainty and stability are not opposites. They are phases of the same structural process.

- When competing strain geometries amplify one another beyond redistribution capacity, articulation occurs. That is instability.
- When competing strain geometries interlock and remain below threshold within extended alignment span, persistence occurs. That is stability.

The difference is not in the existence of strain. It is in how strain is organized.

Stability as Organized Uncertainty

This insight dissolves the apparent contradiction between uncertainty and persistence.

The classical world appears stable not because uncertainty is absent, but because uncertainty is diluted and distributed across scale. Massive aggregation of interactions smooths local fluctuations. Redistribution capacity increases with scale. Alignment spans extend beyond perceptual resolution.

Large systems can carry immense micro-uncertainty without visible articulation.

Atoms vibrate, yet solids persist. Molecules collide, yet materials endure. Cells mutate, yet organisms survive. Individuals conflict, yet societies continue.

Stability is structured tolerance of uncertainty. It is not permanent certainty. It is delayed articulation under controlled strain.

Fragility vs. Flexibility

This also clarifies something profound: fragility is not the presence of uncertainty. It is the inability to organize it.

Rigid systems collapse quickly because they lack redistribution pathways. Flexible systems endure because they integrate distortion dynamically. Stability is therefore dynamic negotiation held below threshold. It is not calm silence. It is disciplined motion.

The Paradox Fully Resolved

We can now state the paradox clearly—and resolve it fully:

- **Uncertainty is structurally inevitable.**
- **Stability is structurally possible.**

Both coexist because stability is not elimination of strain—it is successful management of strain within finite alignment span. Reality sustains not by removing uncertainty, but by distributing it.

Preparing for Scale

This realization prepares us for the next movement. If stability is organized uncertainty, and uncertainty is inevitable, then scale must determine how visible uncertainty becomes.

At microscopic levels, strain density is high and alignment spans are comparable to measurement interaction. Uncertainty appears fundamental.

At macroscopic levels, strain dilution and aggregation extend alignment spans beyond perception. Stability dominates appearance.

The world does not collapse because uncertainty is structured.

And now we must understand how scale transforms its visibility. We move next from persistence to scale and appearance.

From Persistence to Scale

We began this part with a simple but unsettling question: *If uncertainty is inevitable, why does reality not collapse?*

We now have the answer. Reality does not endure because strain disappears. It endures because strain is organized.

Distortions interlock. Local equilibria form. Alignment spans absorb disturbance. Delay extends articulation beyond immediate threshold. Stability is not the elimination of uncertainty. It is the disciplined containment of uncertainty within constraint geometry.

The paradox is resolved. Uncertainty and persistence are not enemies. Persistence is uncertainty managed below threshold.

The Deeper Question

But this resolution introduces a deeper question. If uncertainty is always present beneath stability—why does it appear so dramatically at microscopic scales and so faintly at macroscopic ones?

Why does quantum theory speak openly of indeterminacy, while classical mechanics appears confident and precise? Why do atoms reveal uncertainty, while planets appear predictable?

If stability is organized strain, then scale must determine how strain is distributed, diluted, and perceived.

Scale and Appearance

At small scales, strain density is high. Alignment spans approach structural limits. Competing geometries remain visible to measurement.

At large scales, aggregation smooths micro-competition. Alignment spans exceed observational bandwidth. Uncertainty becomes statistically compressed into apparent determinism.

The difference is not ontological. It is geometric.

The Next Movement

The next movement of this book shifts from what uncertainty is and how stability persists to how scale transforms appearance. We must now examine:

- Why microscopic domains expose indeterminacy openly.
- Why macroscopic domains conceal it beneath apparent certainty.
- How aggregation, density, and bandwidth reshape visibility.

The universe is not divided into deterministic and indeterministic realms. It is continuous. What changes is the way strain organizes across scale.

We now move from persistence to visibility. From stability to scale. And in that transition, the apparent boundary between quantum and classical worlds begins to dissolve.

Interlude — What We Have Discovered So Far

Before we move forward, pause.

If you have read carefully, something subtle has shifted.

Uncertainty no longer feels mystical. Stability no longer feels static. Randomness no longer feels fundamental.

This part was not about adding a new theory. It was about removing a misunderstanding.

Let us gather the structural realizations clearly.

Core Realizations from Part I

1. Nothing Is Random in Ontology

- Randomness implies disconnection.
- In a continuous structure, absolute disconnection is impossible.
- What appears random is unresolved relational geometry.

2. Continuity Is Fundamental

- Reality is not made of isolated fragments.
- Structured continuity underlies all configurations.
- Every local event participates in a larger relational field.

3. Delta Is Geometric Strain

- Disturbance is not an external force.
- Delta is distortion within continuity.
- Strain creates the need for alignment.

4. Stability Is Dynamic Tension

- Stability does not mean absence of motion.
- It means strain is balanced within constraint.
- Persistence is tension organized—not eliminated.

5. Stillness Is a Perceptual Illusion

- What appears static is rapid alignment beyond observational bandwidth.
- Motion never ceases—only its visibility changes.

6. Uncertainty Is Pre-Resolved Strain

- Uncertainty is not ignorance.
- It is distortion that has not yet crossed articulation threshold.
- It exists between strain and resolution.

7. Competition Is Necessary for Uncertainty

- A single viable alignment path produces determinism.
- Multiple viable strain geometries produce indeterminacy.
- Uncertainty requires coexistence of competing configurations.

8. Finite Alignment Threshold Time (ATT) Is Essential

- Resolution is not instantaneous.
- Alignment requires span.
- That span is where uncertainty lives.

9. Bandwidth Is Limited

- Systems process strain within finite capacity.
- Observers perceive within finite capacity.
- Indeterminacy increases when strain complexity approaches bandwidth limits.

10. Threshold Ends Uncertainty

- Resolution is not gradual smoothing.
- It is threshold crossing.
- Articulation occurs when alternatives lose viability.

11. Certainty Is Exhaustion, Not Dominance

- Certainty does not mean absolute truth.
- It means competing geometries are no longer sustainable.
- Certainty is local and temporary.

12. Stability Is Organized Uncertainty

- Persistence does not eliminate strain.
- It redistributes it.
- Stable systems stretch uncertainty without collapsing.

13. Delay Is Structural Strength

- Delay is not weakness.
- It reflects redistribution capacity.
- The ability to sustain pre-resolution span indicates robustness.

14. Local Equilibrium Makes Structure Possible

- Global equilibrium is impossible in finite systems.
- Stability forms in local minima of strain geometry.
- Reality sustains through pockets of negotiated balance.

The Core Shift

Uncertainty is not chaos. Stability is not stillness. Certainty is not dominance. Randomness is not fundamental.

Everything we experience emerges from structured continuity negotiating distortion within finite alignment span.

What Changes Now

If uncertainty is structural and stability is organized strain, then the next question becomes unavoidable:

Why does uncertainty appear more visible at microscopic scales and diluted at macroscopic ones?

Why does the quantum domain seem uncertain, while the classical domain appears deterministic?

This part has given us the architecture. Next part will explore how scale transforms appearance.

We now move from structure to visibility. From inevitability to interpretation. And the world will begin to look very different.

Part II — From Strain to Experience

"Uncertainty does not disappear at scale. It reorganizes. What looks certain is strain diluted beyond visibility."

Overview

The previous part established something foundational:

- Uncertainty is structural.
- Stability is organized strain.
- Certainty is temporary exhaustion.

But if uncertainty is inevitable, a new question emerges:

- Why does the classical world look so stable?
- Why do macroscopic systems behave predictably?
- Why does quantum uncertainty feel confined to microscopic scales?

This part answers by shifting the lens from **existence** to **appearance.**

We now explore how strain behaves across scale:

- How competing geometries dilute through aggregation.
- How alignment spans stretch beyond perception.
- How probability becomes descriptive rather than ontological.
- How collapse is reframed as articulation exhaustion rather

than mysterious jump.

This section clarifies a subtle but decisive insight:

The universe is not divided into deterministic and indeterministic realms. There is only one continuous structure.

What changes is density, bandwidth, interaction, and scale.

By the end of this part, the apparent boundary between quantum uncertainty and classical stability will no longer appear as a philosophical divide—but as a geometric transition.

We now move from **structure** to how structure becomes **experience.**

Chapter 5 — Layers of Reality

When We Confuse Description with Existence

Most paradoxes do not arise from reality. They arise from language.

We describe something, and then we quietly begin to treat the description as the thing itself. An equation becomes an object. A probability becomes a property. A measurement limit becomes a metaphysical boundary. And slowly, confusion hardens into philosophy.

When we say a particle has a probability distribution, what are we really saying? That reality itself is smeared? Or that our description of unresolved alignment must be statistical?

When we say a wavefunction collapses, what collapses? Structure? Or our representation of competing geometries?

When we say uncertainty is fundamental, do we mean structure is indeterminate—or that alignment operates within finite span and limited bandwidth?

The Problem of Description

The problem is subtle. Science gives us extraordinarily powerful models. Mathematics describes behavior with stunning precision. But description is not ontology.

The map is not the terrain. The equation is not the structure. The probability is not the strain.

Yet over time, especially in discussions of quantum mechanics, the layers blur. We begin to speak as if probability waves exist in the same way that atoms exist. We speak as if randomness is woven into the substance of reality itself. We treat epistemic limits—what can be measured—as ontological limits—what can exist. And from that confusion, paradox multiplies.

Restoring Separation

This chapter introduces a discipline. Not a new theory. Not a replacement of physics. A separation.

Reality operates in layers:

- At one layer, there is **structure**—continuity, distortion, geometry.
- At another, there is **relational evolution**—alignment, delay, threshold.
- At another, there is **description**—probability, equations, measurement models.

When we mix these layers, mystery appears where none is required. When we separate them carefully, clarity returns.

Uncertainty becomes structural rather than mystical. Probability becomes descriptive rather than ontological. Collapse becomes articulation rather than magic.

The Core Insight: The universe does not contradict itself. We contradict ourselves when we confuse what *is* with how we describe it.

5.1 — The Ontological Layer (Structured Continuity / SVE)

Before we speak about probability, measurement, collapse, or even time, we must clarify what belongs to the most basic layer: **ontology.**

Ontology answers a simple question: *What exists?* Not what we measure. Not what we calculate. Not what we infer statistically. Simply—what exists.

In the framework developed so far, the ontological layer is defined by **structured continuity**—what we have referred to as **Structured Vacuum Energy (SVE).** This does not introduce a mystical entity. It simply asserts that the substrate of reality is continuous and structured—not empty, not fragmented, not composed of isolated objects floating in void.

At this layer:

- There are no truly isolated particles.
- There are no absolute voids.
- There are no disconnected events.

There are only configurations within continuity.

What we call particles are localized configurations. What we call fields are structured regions. What we call objects are stabilized strain geometries. But beneath all of them, continuity remains.

No Probability at the Ontological Layer

This has an important implication: at the ontological layer, there is **no probability.**

Probability does not exist as a physical substance. It is not woven into the fabric of reality. There is no "probability field" oscillating in space. There is no randomness stored in structure itself.

There is geometry. There is distortion (*delta*). There is tension. There is alignment in progress. But there is no randomness as a primitive entity.

This distinction is essential.

When we say a system has multiple possible outcomes, that statement belongs to the **epistemic layer.** At the ontological layer, there are competing strain geometries—not abstract possibilities floating without structure. Possibility is our word. Strain geometry is the structure's condition.

Similarly, the wavefunction—so central in quantum mechanics—does not belong to ontology in the sense of being a literal object in space. It belongs to the **descriptive layer.** Ontologically, what exists is structured continuity negotiating distortion. The wavefunction is a representation of unresolved strain geometry. Confusion arises when we elevate representation into substance.

Ontology Is Austere

Ontology is austere. It contains:

- Continuity.
- Geometric distortion.
- Constraint relations.
- Threshold conditions.

Nothing more.

No equations. No probabilities. No measurement errors. No interpretive overlays. Those belong to other layers.

This discipline matters because many philosophical puzzles arise from **ontological inflation**—when we treat descriptive constructs as if they physically exist in the same way as structure.

If probability is mistaken for an ontological ingredient, then randomness becomes fundamental. If wavefunction collapse is mistaken for a physical tearing of reality, then mysticism creeps into physics.

But if we remain disciplined, the ontological layer is quiet. It contains only structured continuity and its distortions.

At this level:

- Nothing is random.
- Nothing is probabilistic.
- Nothing collapses mysteriously.

There is only strain geometry evolving under constraint.

Uncertainty does not live here as ignorance or chance. It exists here as pre-resolved distortion—not as statistical ambiguity.

The ontological layer is silent about knowledge. It simply contains structure. And because it contains structure, it cannot contain absolute disconnection. Randomness, in the sense of causeless isolation, cannot exist at this layer.

Everything that follows—alignment span, threshold articulation, probability distributions—emerges from this structured base. If we lose clarity at this foundational layer, all higher interpretations become unstable.

So we anchor firmly: **Ontology = structured continuity under distortion. Nothing more.**

With ontology cleanly defined, we can now move to the second layer: not what exists, but how it evolves. That is the **relational layer.**

5.2 — The Relational Layer (ATT and Alignment)

If the ontological layer answers *what exists*, the relational layer answers a different question: *how does it evolve?*

Ontology gives us structured continuity and geometric distortion. Relational dynamics tell us how distortion propagates, negotiates, and articulates within that continuity.

This is where Alignment Threshold Time (ATT) belongs.

ATT does not exist as a separate substance. It is not a dimension floating beside space. It is not an independent entity. It is **relational span**—the structural duration required for distortion to propagate, compete, and cross articulation threshold.

The ontological layer contains strain. The relational layer contains negotiation.

Relational Dynamics

When delta (geometric strain) arises in structured continuity, it does not resolve everywhere at once. Alignment must propagate through constraint geometry. Competing distortions interact. Redistribution pathways engage. Threshold conditions are approached.

This process unfolds relationally. Relational dynamics include:

- Propagation speed of alignment.
- Interaction strength between competing geometries.
- Redistribution capacity.
- Threshold sensitivity.

These are not ontological substances—they are properties of how structure behaves under distortion.

Where Uncertainty Lives

Uncertainty, as defined earlier, lives primarily in this relational layer.

At the ontological level, there is simply strain. At the relational level, that strain competes within finite alignment span.

Indeterminacy arises not because ontology is ambiguous, but because relational negotiation is incomplete.

This distinction is critical.

When we speak of **superposition** in quantum mechanics, we often imagine something mysterious existing in reality itself. But structurally, superposition belongs to the relational layer: it reflects coexistence of competing strain geometries prior to articulation.

When we speak of **collapse**, we are describing a relational event: threshold crossing that exhausts competing viability. Collapse is not an ontological rupture. It is a relational articulation.

Functions of the Relational Layer

The relational layer governs:

- Delay between disturbance and outcome.
- Formation of local equilibrium.
- Redistribution of strain.
- Stability margins.
- Threshold transitions.

It explains why certainty appears sudden—because articulation compresses negotiation into visible event. It explains why stability persists—because ripple-lock geometries redistribute distortion below threshold. It explains why uncertainty exists—because competing geometries coexist within finite alignment span.

Most importantly, it explains why indeterminacy does not contradict structure. Indeterminacy arises in the relational layer, not in ontology. Ontology contains geometry. Relational dynamics determine how that geometry evolves.

Avoiding Confusion

Confusion occurs when we attribute relational properties to ontology. For example:

- Saying *"reality is probabilistic"* mistakes epistemic description for relational delay.
- Saying *"the wavefunction collapses physically"* mistakes relational articulation for ontological discontinuity.

The relational layer is where time, threshold, and interaction reside. It is dynamic. But it remains fully structural. Nothing mystical enters here. Nothing random is injected.

Finite alignment span guarantees that multiple strain geometries can coexist temporarily. That coexistence produces uncertainty in the relational layer.

The ontological layer remains continuous. The relational layer negotiates distortion. Keeping these layers distinct prevents confusion.

With ontology defined and relational dynamics clarified, we now move to the third layer—the one most familiar in science: *how we describe what we observe*. That is the *epistemic layer.*

5.3 — The Epistemic Layer (Probability and Description)

If the ontological layer answers *what exists* and the relational layer explains *how it evolves*, the epistemic layer answers a different question: *how do we describe it?*

This is the layer of models, equations, probabilities, measurements, error bars, and wavefunctions. It is powerful. It is indispensable. But it is not ontology.

The epistemic layer does not create reality. It organizes our access to it.

Probability as Description

When we assign a probability to an outcome, we are not inserting randomness into structure. We are describing the distribution of viable articulations within unresolved relational dynamics. Probability belongs to description.

If competing strain geometries coexist within finite alignment span, and if we do not know which one will exhaust alternatives first, we assign probabilities. Those probabilities reflect geometric viability, interaction strengths, boundary conditions, and observational limits. They do not imply ontological dice.

Similarly, the **wavefunction** in quantum mechanics is a descriptive tool. It encodes the relational configuration space of possible articulations under current constraints. It maps unresolved geometry. It does not represent a physical mist floating in space.

When measurement occurs, and the wavefunction "collapses," what actually collapses is the descriptive representation of competing configurations. Ontologically, strain geometry articulates through threshold. Relationally, alignment completes. Epistemically, the probability distribution updates. Collapse is epistemic compression of relational articulation.

Dissolving Confusion

This distinction dissolves much confusion. We often hear statements such as:

- "Reality is probabilistic."
- "The particle exists in multiple states simultaneously."

- "The wavefunction is real."

But these claims blur layers. Probability is epistemic. Superposition is relational. Structure is ontological.

At the epistemic layer, uncertainty appears as error margins, statistical spreads, Heisenberg inequalities, and probabilistic amplitudes. These tools quantify limits of prediction and measurement under finite alignment span and limited bandwidth. They are extraordinarily successful. But they describe—they do not constitute.

The **Heisenberg uncertainty principle**, for example, does not mean that reality is fundamentally fuzzy in its being. It means that within relational dynamics and finite interaction, precise simultaneous articulation of certain variables is structurally constrained. Our description must reflect that constraint statistically.

The Discipline of Description

The epistemic layer is shaped by both ontology and relational dynamics.

- Because alignment is finite, we cannot predict with absolute certainty.
- Because bandwidth is limited, we cannot measure with infinite precision.
- Because competition exists, we must assign probabilities to articulation outcomes.

All of this belongs to description.

The danger arises when description is mistaken for existence. When probability is treated as substance, randomness becomes metaphysical. When equations are treated as objects, models are confused with structure.

The epistemic layer must remain disciplined. It is the map, not the terrain.

This does not weaken science. It strengthens interpretation. It prevents philosophical inflation. The wavefunction remains valid. Probability remains essential. Statistical mechanics remains powerful. But they operate at the epistemic layer.

They describe unresolved relational dynamics grounded in structured ontology.

Clarity Restored

When we keep this separation clear, paradox diminishes. Quantum indeterminacy becomes relational negotiation under finite alignment span. Probability becomes geometric viability distribution. Collapse becomes articulation threshold crossing. Mystery yields to structure.

The final task in this chapter is to diagnose why confusion between these layers happens so persistently. That is where we now turn.

5.4 — Why Confusion Happens Between Layers

If the separation between layers is so clarifying, why are they so often mixed? Why do we repeatedly mistake description for existence, relational delay for randomness, and mathematical representation for physical substance?

The confusion is not accidental. It arises naturally from how we experience and communicate reality.

1. Success of Mathematics

Mathematics is extraordinarily successful. Equations predict with astonishing precision. Probability distributions match experiments. The wavefunction produces correct results. When a tool works so well, we instinctively elevate it. We begin to speak of it not merely as a description, but as reality itself.

Success breeds ontological inflation. The map becomes so accurate that we forget it is a map.

2. Compression Through Language

Language compresses layers automatically. When we say, "*The electron is in a superposition,*" we are using epistemic shorthand for a relational condition. But grammar makes it sound ontological. The structure of language tempts us to treat relational states as intrinsic properties.

We say, "*The wavefunction collapses,*" instead of, "*Our descriptive model updates after relational articulation.*" Language smooths nuance. Philosophy then hardens the smoothed version into doctrine.

3. Measurement as Interaction

Measurement is interaction. In quantum systems especially, the act of measurement alters relational dynamics. Because observation influences articulation, it becomes tempting to conclude that knowledge creates reality.

But measurement belongs to the relational layer—it is an interaction that constrains geometry. The epistemic update that follows is not ontological creation. It is descriptive revision. Yet because measurement and knowledge occur together in experience, the layers blur psychologically.

4. Probability Feels Real

Repeated exposure to statistical distributions—radioactive decay rates, interference patterns, thermal fluctuations—makes probability appear as a physical ingredient. It seems woven into reality.

But probability emerges because relational dynamics under finite alignment span produce outcome distributions across repeated trials. The distribution reflects geometry and constraint, not ontological randomness. Our repeated exposure to statistical patterns conditions us to treat probability as substance.

5. Scale Amplifies Confusion

At macroscopic scales, determinism appears dominant. At microscopic scales, probability dominates description. Without layer discipline, this contrast appears metaphysical. It seems as though reality itself changes character across scale.

But what changes is visibility of relational negotiation relative to bandwidth. When we mix layers, quantum mechanics looks mystical. When we separate them, it becomes structurally coherent.

Interpretive Consequences

Most philosophical debates about quantum interpretation—Copenhagen, many-worlds, objective collapse, hidden variables—are attempts to reconcile ontological and epistemic statements without explicit layer separation.

The disagreement often arises not from data, but from layer blending. Some interpretations treat the wavefunction as ontological. Others treat it as epistemic. Others mix relational and ontological claims. Without clear boundaries, paradox multiplies.

Why We Blur Layers

The confusion is understandable. Human cognition prefers unified narratives. We do not naturally think in layered frameworks. We compress structure, behavior, and description into a single story.

But clarity requires discipline. Ontology tells us what exists. Relational dynamics tell us how it evolves. Epistemic tools tell us how we describe it.

When we mistake one for another, mystery appears. When we keep them separate, structure emerges cleanly.

The paradox of uncertainty dissolves not because reality changes—but because interpretation becomes layered.

Toward Methodology

The final step in this chapter is not merely diagnostic. It is methodological. We must establish a discipline of separation going forward. That is where we now turn.

5.5 — The Discipline of Separation

We have identified the layers. We have diagnosed the confusion. Now we establish the rule that will govern everything that follows.

Separation is not optional. It is methodological discipline.

If we want clarity about uncertainty—especially in quantum mechanics, probability theory, and large-scale stability—we must never collapse ontology, relational dynamics, and epistemic description into one undifferentiated narrative.

The Rules of Discipline

Never Elevate Description into Substance

- Equations are tools. Wavefunctions are representations. Probability distributions are summaries of viability.
- They are not ontological objects.
- When we treat mathematical constructs as physical entities, paradox arises immediately. A wavefunction does not "live" in space the way matter does. It encodes relational possibilities under constraint.
- Keep the map as a map.

Never Treat Relational Delay as Randomness

- Finite alignment span allows competing strain geometries to coexist. That coexistence produces indeterminacy at the relational level.
- But relational indeterminacy is not ontological chaos. It is structured negotiation under constraint.
- Randomness is a descriptive shorthand for unresolved relational competition. Do not turn shorthand into metaphysics.

Never Confuse Measurement Limits with Reality Limits

- Measurement constrains geometry. Observation updates description.
- But the fact that we cannot measure something simultaneously with infinite precision does not mean structure lacks coherence.
- The Heisenberg uncertainty principle describes relational constraints under measurement interaction. It does not declare reality inherently irrational.

- Bandwidth limitation is structural—not mystical.

Always Ask: Which Layer Does This Claim Belong To?

- When encountering a statement like:
    - "Reality is probabilistic."
    - "The particle exists in multiple states."
    - "Collapse creates reality."
- Pause. Ask: Is this ontological? Is this relational? Or is this epistemic?
- Many contradictions vanish the moment the layer is identified.

Reconstruct the Full Chain

- Whenever confusion arises, rebuild the hierarchy:
    - **Ontology** → **Structure** (continuity, strain geometry).
    - **Relational Dynamics** → **Alignment** (ATT, competition, threshold).
    - **Epistemic Representation** → **Description** (probability, equations, models).
- Keep this order intact. Reality does not begin with equations. Equations arise because structure behaves relationally under constraint.

Why Discipline Matters

This discipline does not weaken science. It strengthens it. It prevents philosophical inflation. It prevents mystification. It prevents unnecessary metaphysical battles.

Most importantly, it allows us to reinterpret quantum mechanics without discarding its mathematics. The wavefunction remains valid.

Probability remains indispensable. Statistical mechanics remains powerful. But they remain epistemic tools describing relational dynamics grounded in structured ontology.

The Remarkable Consequence

Once separation is disciplined, something remarkable happens: the boundary between classical and quantum begins to soften. What looked like two different realities begins to appear as one structure seen at different scales and bandwidths.

And now we are ready to take the next step. With layers clearly separated, we can return to the heart of the mystery: *superposition, collapse, and measurement*. Not as metaphysical puzzles, but as structured relational events described epistemically.

We now move from layer clarity to quantum reinterpretation.

From Layer Discipline to Quantum Clarity

We have done something deceptively simple in this chapter. We have separated *what is*, from *how it evolves*, from *how we describe it.* And in doing so, much of the fog begins to lift.

Ontology contains structure. Relational dynamics contain alignment and threshold. Epistemic tools contain probability and equations.

When these remain distinct, uncertainty is no longer mysterious. It becomes structural. When they collapse into one narrative, paradox multiplies.

Why Discipline Matters

It is worth pausing here. Many of the deepest debates in modern physics do not arise from experimental contradiction. They arise from interpretive blending. We argue about whether the wavefunction is "real," whether collapse is "physical," whether probability is "fundamental," without first clarifying which layer we are speaking from.

Layer confusion creates metaphysical tension where structural coherence already exists.

Reinterpreting Quantum Mechanics

Now that discipline is restored, we are prepared for something important. We can revisit quantum mechanics—not to replace it, not to challenge its mathematics—but to reinterpret it cleanly.

- **Superposition** will no longer mean "a particle literally in two contradictory states."
- **Collapse** will no longer require metaphysical rupture.
- **Probability** will no longer imply ontological randomness.

Instead, we will see:

- Superposition as relational coexistence of competing strain geometries.
- Collapse as articulation through threshold exhaustion.
- Probability as epistemic distribution of viability within unresolved alignment span.

Mystery Dissolved

The mystery will not vanish because we deny it. It will dissolve because we understand its layer.

From here onward, we move deeper into the quantum domain—not to mystify it, but to place it correctly within structured continuity.

The next chapter enters directly into *superposition and collapse.* And this time, we will carry our discipline with us.

Chapter 6 — The Geometry of Delay

The Myth of Instant Resolution

We speak as if events happen in an instant. The glass shatters. The particle collapses. The verdict is delivered. The decision is made.

Language compresses reality into sharp moments. But what if nothing truly happens instantaneously? What if every visible event is the final articulation of a geometric negotiation that has been unfolding beneath perception?

Hidden Preparation

When a crack appears in a material, we call it sudden. Yet long before the visible fracture, microscopic strain accumulated, redistributed, concentrated, and approached threshold. The articulation looks abrupt. The preparation was gradual.

When a person makes a decision, it feels like a moment of clarity. But competing narratives, emotional tensions, and cognitive strain have been negotiating internally. The final "yes" or "no" is threshold crossing—not the beginning of process.

Even in quantum mechanics, we speak of wavefunction collapse as if it were an instantaneous leap. But is it a leap? Or is it the articulation of relational strain under constraint?

The illusion of instant resolution arises because articulation compresses prior negotiation into a single perceptible event. Threshold crossing hides history.

Time as Negotiation

This illusion has shaped how we think about time itself. We imagine events occurring in discrete instants, as if reality moves from one frozen frame to another. But if alignment requires propagation, and propagation requires finite span, then no articulation is truly instantaneous.

Every event has geometry. Every articulation has preparation. Every "moment" contains accumulated alignment span that preceded it.

What we perceive as an instant is the limit of our bandwidth. We do not see the relational unfolding below threshold. We only see the crossing.

Delay as Structure

This misunderstanding is not trivial. If we assume resolution is instantaneous, we imagine time as a container in which events simply appear. But if resolution requires span, then time is not a passive backdrop. It is the measure of relational negotiation.

Delay is not waiting. It is structure unfolding.

And once we recognize that nothing resolves instantly, a deeper insight emerges:

- Uncertainty exists because resolution takes span.
- Time exists because alignment is finite.
- Irreversibility exists because threshold changes geometry.

ATT as Necessity

In this chapter, we will examine delay not as inconvenience, but as structural necessity. We will see that Alignment Threshold Time

(ATT) is not an abstract parameter—it is the geometric duration of resolution itself.

And once we understand the geometry of delay, the arrow of time will no longer appear mysterious. It will appear inevitable.

We begin by dismantling the illusion of instant resolution.

6.1 — Resolution Is Not Instantaneous

The appearance of suddenness is one of the most persistent illusions in our experience of reality.

We see the glass shatter and assume the break happened in a single moment. We observe a phase transition and imagine the system "jumped." We measure a quantum state and describe collapse as instantaneous.

But articulation is not the beginning of a process. It is the end of one.

Negotiation vs. Articulation

To understand this, we must distinguish between *negotiation* and *articulation*.

- Negotiation unfolds below threshold.
- Articulation occurs at threshold.

Most of what determines an event happens before it becomes visible.

Take material fracture. A metal rod under increasing load does not fail at the first application of force. Micro-strains accumulate. Dislocations propagate. Internal stress redistributes unevenly. Competing micro-cracks form and vanish. The structure negotiates distortion.

For a long time, nothing visible happens. Then, at a critical concentration of strain, one crack geometry exhausts alternatives. Redistribution capacity fails. Threshold is crossed.

The fracture appears sudden. But the suddenness is perceptual compression. The geometry had been evolving continuously.

Cognitive Decisions

The same logic applies to cognition. When someone says, *"I've made up my mind,"* it feels like a discrete decision. But relationally, competing narratives have been interacting. Emotional strain, memory, anticipation, identity—all contribute to internal negotiation.

The articulation—the spoken decision—is threshold crossing. Resolution is rarely instantaneous; articulation is merely sharp.

Quantum Measurement

Now consider quantum measurement. In standard interpretation, wavefunction collapse is often described as instantaneous. A system in superposition becomes a definite state in a flash.

But relationally, measurement is interaction. Interaction constrains geometry. Competing strain configurations interact with apparatus constraints. Viability shifts. Redistribution pathways narrow.

When threshold is reached, only one configuration remains sustainable. The collapse is not an ontological jump. It is relational exhaustion under constraint.

The articulation may appear instantaneous relative to measurement resolution, but alignment required span.

Why This Matters

This distinction matters deeply. If we mistake articulation for process, we believe resolution happens without geometry. That illusion leads to metaphysical speculation.

But once we see that:

- Every articulation is preceded by negotiation.
- Every negotiation requires finite span.
- Every span reflects propagation within constraint.

Then the myth of instant resolution dissolves.

Threshold Compression

Even in macroscopic systems, where determinism appears dominant, delay exists. Chemical reactions have activation pathways. Mechanical failures have fatigue accumulation. Political revolutions have long periods of tension before eruption.

What appears sudden is structurally prepared. The threshold compresses history.

This compression is especially powerful because human perception is threshold-oriented. We detect change most strongly at articulation points—when something crosses from one state to another. We rarely perceive the continuous negotiation leading up to it.

Thus, our intuition tells us: *events happen suddenly.* Structure tells us: *events articulate after span.*

Time as Propagation

This insight reframes how we understand time. If resolution is never instantaneous, then every event has geometric duration. That duration is not merely clock time. It is relational propagation.

And that propagation is what we call Alignment Threshold Time (ATT).

We now formalize this span more precisely.

6.2 — ATT as Structural Duration

If resolution is never instantaneous, then the span between disturbance and articulation must be understood structurally. That span is what we call Alignment Threshold Time (ATT).

But we must be careful. ATT is not clock time. It is not an independent physical substance. It is not a separate dimension inserted into reality.

ATT is structural duration.

It is the geometric span required for distortion to propagate, compete, redistribute, and cross articulation threshold within constraint geometry.

How ATT Operates

When delta arises—when strain appears in structured continuity—it does not resolve everywhere at once. Alignment must travel through relational pathways. Competing geometries must negotiate. Redistribution channels must engage.

The duration of this negotiation is ATT.

- In fragile systems, alignment span is short. Strain concentrates quickly. Threshold is crossed rapidly.
- In robust systems, alignment span is extended. Strain

redistributes widely. Threshold crossing is delayed or avoided.

ATT therefore reflects structural capacity.
It depends on:

- Constraint geometry.
- Redistribution pathways.
- Interaction density.
- Strain magnitude.
- Bandwidth limits.

It is not universal in value, but universal in principle. Every articulation—whether quantum collapse, mechanical fracture, chemical reaction, cognitive decision, or social upheaval—has an associated ATT.

Why Events Appear Instantaneous

The appearance of instant resolution often arises because ATT is small relative to our perceptual bandwidth. If alignment completes faster than we can resolve, it appears instantaneous. But structural duration still existed.

This distinction becomes critical in quantum mechanics. When we say a state collapses instantly upon measurement, what we often mean is that collapse occurs within interaction timescales too small for classical measurement resolution.

But interaction always requires span. Alignment always propagates. Constraint always reshapes geometry before articulation. ATT does not disappear at small scales—it becomes comparable to interaction scales.

Quantum vs. Classical Scales

This explains why quantum systems display indeterminacy so prominently.

At microscopic scales:

- Alignment spans are small but not zero.
- Interaction densities are high.
- Competing geometries coexist within tight bandwidth limits.

Uncertainty becomes visible because ATT is not negligible relative to observation.

At macroscopic scales:

- Alignment spans extend across large redistribution networks.
- Aggregation smooths micro-negotiations.
- Bandwidth dilution hides micro-competition.

ATT becomes embedded in classical time flow.

Time as Alignment Span

Time, then, is not merely a parameter. It is accumulated alignment span. The ticking of a clock does not create duration. It measures cyclic articulation within structured negotiation.

This reframes time fundamentally. Time is not something in which events occur. Events occur because alignment requires span.

ATT is not an abstract addition to physics. It is recognition that structure cannot resolve without duration.

The Core Conclusion

And this leads to a powerful conclusion:

- Without ATT, uncertainty cannot exist.
- Without ATT, stability cannot form.
- Without ATT, time itself has no meaning.

Resolution requires propagation. Propagation requires geometry. Geometry unfolding requires span. ATT is that span measured structurally.

Now we extend this insight further. If ATT is structural duration, then what we experience as time may be nothing more than perceived alignment span.

We turn now to that question directly.

6.3 — Time as Perceived Alignment Span

If ATT is structural duration, then what we call time must be reconsidered.

We are accustomed to thinking of time as a container—an invisible river in which events float. Seconds pass. Minutes accumulate. History unfolds inside an independent dimension.

But what if time is not the container? What if time is the measure of alignment unfolding?

Alignment Produces Time

If distortion requires propagation, and propagation requires span, then duration is not an external stage. It is the geometry of resolution itself.

Alignment does not occur *in* time. Alignment produces time.

When nothing changes—when no distortion propagates—time feels absent. In deep sleep, under anesthesia, or in moments of suspended awareness, subjective time collapses. When strain density increases—during crisis, intense focus, or high emotional tension—time stretches or compresses in perception.

Why? Because perceived time correlates with alignment density. Where negotiation is dense, duration feels extended. Where negotiation is sparse, duration feels minimal.

This suggests something fundamental: *time is not merely a physical coordinate. It is experienced alignment span.*

Layers of Time

- At the **relational layer**, ATT measures structural duration.
- At the **epistemic layer**, clocks measure repeating articulations.
- At the **experiential layer**, consciousness registers alignment flow.

These are different expressions of the same phenomenon.

Consider a mechanical clock. Its ticking is not time itself. It is periodic articulation of mechanical strain within a constrained geometry. We call the interval between ticks "one second," but what we are really measuring is a stable alignment cycle.

The clock does not generate time. It encodes structured delay.

Similarly, in physics, time parameters in equations track evolution of state variables under relational dynamics. But those parameters are abstractions of alignment span.

If alignment were instantaneous—if ATT were zero—there would be no duration. No evolution. No sequence. Events would not unfold. They would exist in static simultaneity.

Finite ATT makes sequence possible. Sequence produces causality. Causality produces the perception of time.

The Arrow of Time

This reframes the arrow of time as well. We do not move through time like objects moving through space. Rather, alignment unfolds through structured continuity. The sequence of articulations creates ordered experience.

Time is structured progression of resolution.

This explains why time feels directional. Once articulation exhausts competing geometries, structure changes irreversibly under given constraints. Reconstruction requires reintroducing prior strain configurations—a process that itself requires alignment span and redistribution.

Directionality emerges from threshold exhaustion. Time does not flow because of a metaphysical arrow. It flows because geometry unfolds irreversibly once articulation occurs.

Scale and Continuity

At microscopic scales, alignment spans are small and relational competition is dense. Duration becomes less intuitive. Quantum processes appear discontinuous because ATT operates near interaction scales.

At macroscopic scales, aggregated alignment spans produce smooth temporal continuity. Classical time emerges as averaged relational unfolding.

In this light, time is not an independent dimension sitting beside space. It is the perceptible trace of alignment in motion.

Without strain, no alignment. Without alignment, no span. Without span, no time.

Time is the experiential signature of structured delay.

And once we recognize time as perceived alignment span, the final question becomes unavoidable: Why does alignment produce irreversibility? Why does the past feel fixed and the future open?

The answer lies in how threshold changes geometry.

We now turn to *irreversibility*.

6.4 — Why Irreversibility Appears

If time is alignment unfolding, then irreversibility must be understood structurally—not philosophically.

Why does the past feel fixed? Why do certain processes move in one direction? Why does a shattered glass not spontaneously reassemble? Why does articulation seem to close possibilities rather than preserve them?

The answer lies in threshold geometry.

Threshold Crossing

Before articulation, multiple strain geometries coexist within finite alignment span. This coexistence is what we call uncertainty. Competing configurations remain viable. The system is negotiating.

Once threshold is crossed, one geometry exhausts alternatives. Competing strain configurations lose viability. The structure reorganizes around the articulated state. That reorganization changes constraint geometry.

Irreversibility appears because threshold crossing modifies the landscape.

Examples Across Domains

- **Fracture:** Before a crack forms, strain is distributed. Multiple micro-crack pathways are viable. Once a crack propagates beyond critical length, the material's geometry changes permanently. Stress redistributes differently. Reconstruction would require reintroducing precisely the prior strain distribution—a configuration that no longer exists naturally. The path is not forbidden by metaphysics. It is improbable because geometry has changed.

- **Entropy:** In thermodynamics, entropy is not disorder in a mystical sense. It is distributed strain across many degrees of freedom. As alignment propagates and articulation occurs, strain spreads into increasingly numerous micro-configurations. To reverse the process, one would need to compress distributed strain back into its prior concentrated geometry. That requires coordinated alignment across enormous degrees of freedom—structurally possible in principle, but overwhelmingly improbable.
- **Cognition:** Before a decision, multiple narratives coexist. After articulation, identity reconfigures. The prior mental configuration cannot simply be restored without reintroducing earlier relational tensions. Memory changes structure. Experience changes geometry.

Structural Consequence

Irreversibility is not a fundamental arrow imposed externally. It is the structural consequence of articulation altering constraint geometry. Once threshold is crossed, the viability landscape shifts. The system cannot return without reconstructing lost relational configurations—an increasingly improbable task as redistribution spreads.

This also explains why microscopic processes sometimes appear reversible while macroscopic processes do not.

- At small scales, alignment spans are short, and the number of degrees of freedom involved in redistribution is limited. Reversal of micro-states is sometimes feasible under controlled conditions.
- At macroscopic scales, articulation involves vast networks of relational degrees of freedom. Redistribution spreads across large geometries. Reversing that spread becomes practically impossible.

Irreversibility scales with complexity of redistribution.

The Arrow of Time

The arrow of time is therefore not mystical. It is geometric.

- Uncertainty exists before threshold.
- Delay measures alignment span.
- Irreversibility emerges after threshold crossing.

Time's arrow is not an independent force. It is the structural trace of articulated strain.

Toward Synthesis

Irreversibility is exhaustion made visible. Once one geometry stabilizes and alternatives lose viability, the system moves forward—not because time pushes it, but because geometry has changed.

This insight unifies uncertainty, delay, and irreversibility. Together, they reveal time as the structured progression of strain through negotiation, articulation, and redistribution.

We now gather these threads into a final synthesis—connecting delay, uncertainty, and time into a single geometric framework.

6.5 — Delay as the Bridge Between Uncertainty and Time

We can now gather the threads.

Uncertainty exists because resolution requires span. Time exists because alignment unfolds. Irreversibility exists because threshold alters geometry.

The bridge between all three is *delay*.

Delay as Structured Negotiation

Delay is not passive waiting. It is structured negotiation under constraint.

Before articulation, multiple strain geometries coexist. That coexistence is uncertainty. It lives inside alignment span. If alignment were instantaneous, competing configurations would never coexist. Uncertainty would vanish.

- No delay → no uncertainty.

But if alignment requires span, then sequence becomes inevitable. One articulation follows another. Each threshold crossing modifies geometry. That sequence is what we perceive as temporal flow.

- No delay → no sequence.
- No sequence → no time.

Delay Across Layers

Delay therefore connects ontology and experience.

- At the **ontological layer**, distortion exists as strain

geometry.

- At the **relational layer**, alignment propagates over finite span (ATT).
- At the **epistemic layer**, we measure duration and assign time coordinates.

Time is not separate from delay. It is delay accumulated across articulations. Every articulation consumes span. Every span contributes to perceived duration.

Uncertainty Generates Time

This reframes the entire relationship between uncertainty and time.

Uncertainty is not something that happens "in" time. Uncertainty is the structural condition that generates time.

As long as competing geometries coexist, delay exists. As long as delay exists, articulation is deferred. That deferral creates sequence. Sequence produces causality. Causality produces direction. Direction produces the arrow of time.

Time is the experiential trace of unresolved strain negotiating toward articulation.

Stability and Fragility

- Stable systems stretch delay without crossing threshold. That is why stable systems persist in time.
- Fragile systems compress delay and articulate rapidly. That is why breakdown feels sudden.

Even at quantum scales, what appears as probabilistic collapse is structured delay resolving under constraint. Even at cosmic scales, stellar evolution, planetary motion, and thermodynamic expansion are alignment processes unfolding across immense spans.

Delay as Universal

Delay is universal. It operates at every scale because finite alignment is universal.

This insight unifies what often appears separate:

- Quantum indeterminacy
- Classical duration
- Entropic irreversibility
- Psychological time perception

All emerge from the same geometry of delay.

Without delay, the universe would be static simultaneity. Without delay, no event could unfold. Without delay, no articulation could distinguish before from after.

Delay is not weakness in structure. It is the condition that makes evolution possible.

Uncertainty stretches delay. Threshold ends delay. Time measures delay.

Toward Scale

And with this, we have reinterpreted time not as a mysterious dimension, but as structured alignment in motion.

The next question becomes unavoidable: If delay operates universally, and if time emerges from alignment span, then how does scale transform this delay into the appearance of determinism at large scales and indeterminacy at small scales?

We now move toward *scale and aggregation*—where delay becomes either visible uncertainty or hidden determinism.

From Delay to Determinism

We began this chapter by questioning something simple: *Is anything truly instantaneous?*

Now we have a structural answer.

Resolution is not a jump. It is alignment unfolding. Articulation is not a beginning. It is threshold crossing after negotiation. Time is not a container. It is accumulated alignment span. Irreversibility is not mystical. It is geometric exhaustion.

And at the center of all of it lies **delay.**

Delay allows competing geometries to coexist. Delay stretches uncertainty. Delay creates sequence. Sequence creates direction.

Without delay, there is no uncertainty. Without delay, there is no time. Without delay, there is no evolution.

The Paradox of Scale

But this realization now confronts us with a deeper paradox.

If delay is universal, if uncertainty is structural, if alignment always requires span—why does the macroscopic world appear so cleanly deterministic?

Why do planets follow predictable orbits? Why do classical equations describe large-scale motion with such precision? Why does everyday reality feel stable and causal?

At microscopic scales, delay manifests as visible indeterminacy. At macroscopic scales, delay seems to dissolve into smooth continuity.

Has uncertainty vanished? Has alignment become instantaneous? Has geometry simplified?

Or has delay transformed through aggregation?

Delay Reorganized

The answer cannot contradict what we have already established. Delay does not disappear. It reorganizes.

As scale increases, redistribution pathways multiply. Competing geometries aggregate. Micro-negotiations average into macro-regularity.

What appears deterministic may simply be uncertainty diluted beyond perceptual resolution.

Toward Scale and Aggregation

The next movement of this book explores that transformation:

- How scale reshapes delay.
- How aggregation compresses indeterminacy.
- How classical determinism emerges from microscopic uncertainty.

We now move from the geometry of delay to the architecture of scale. From uncertainty visible to uncertainty hidden.

And in that transition, the apparent divide between quantum and classical worlds begins to dissolve.

Chapter 7 — Collapse Reframed

The Moment That Divided Physics

Few concepts in modern science have generated as much philosophical tension as *collapse.*

A system is described by a wavefunction. It exists in superposition. Measurement occurs. The wavefunction collapses. One outcome remains.

The language is dramatic. It sounds as if reality hesitates, then suddenly decides. It suggests a discontinuity in the fabric of existence—something multiple becoming singular, something spread out becoming localized.

This single idea has divided physics for decades. Does collapse physically occur? Is it merely epistemic? Does consciousness trigger it? Does it imply multiple worlds? Entire interpretations of quantum mechanics stand or fall on how collapse is understood.

But before entering interpretations, we must pause and ask a more fundamental question: *What if collapse is not what we think it is?*

Collapse Through Layer Discipline

Earlier chapters established three disciplines:

- Ontology contains structured continuity and strain geometry.
- Relational dynamics govern alignment and threshold.

- Epistemic tools describe unresolved viability through probability.

If we apply that discipline here, collapse cannot be an ontological tear in continuity. Structured continuity does not rupture simply because measurement occurs.

Nor can collapse be mystical randomness suddenly injected into structure. Strain geometry evolves under constraint; it does not flip arbitrarily.

So what is collapse?

Collapse as Exhaustion

Perhaps collapse is not a jump. Perhaps it is exhaustion.

Before measurement, multiple strain geometries coexist within finite alignment span. They are viable configurations under current constraints. This coexistence is what we call *superposition* at the epistemic layer.

When measurement interaction occurs, constraint geometry changes. Redistribution pathways narrow. Competing configurations lose viability at different rates. One geometry survives threshold; others exhaust.

The moment we call *collapse* may simply be the articulation of one geometry after alternatives become unsustainable.

If that is true, then collapse is not the birth of reality. It is the end of competition.

The Drama Dissolves

What seemed like a metaphysical leap becomes structural threshold crossing. And perhaps the reason collapse feels so decisive is not because reality jumps—but because articulation compresses negotiation into a single perceptible event.

We only see the crossing. We do not see the exhaustion.

Collapse Reframed

This chapter reframes collapse completely:

- Not as a jump in ontology.
- Not as a magical intervention.
- Not as consciousness creating reality.

But as *relational articulation after geometric exhaustion.*

And once collapse is reframed this way, the divide between quantum and classical begins to narrow. Because articulation exhaustion is not unique to quantum systems. It is universal.

We begin by removing the myth of the jump.

7.1 — Not a Jump, but Articulation Exhaustion

The traditional image of collapse is dramatic. A system is described by a spread of possibilities. Measurement happens. Instantly, one possibility becomes real. It feels like a jump in existence.

But let us slow the narrative down and apply structural discipline.

What Exists Before Measurement

At the **ontological layer**: structured continuity under strain.

At the **relational layer**: competing strain geometries negotiating within finite alignment span.

At the **epistemic layer**: a probability distribution representing viable articulations.

Superposition, then, is not a particle being in contradictory states. It is relational coexistence of multiple viable strain geometries under current constraints. Nothing mystical—just unresolved competition.

Measurement as Coupling

Measurement is not a magical act. It is coupling: a physical interaction between systems. When a measuring apparatus interacts with a quantum system, constraint geometry changes.

Coupling reshapes relational dynamics. Competing geometries that were previously viable now experience differential constraint. Some redistribute strain more efficiently under the new geometry. Others cannot sustain themselves.

As alignment propagates, viability shifts. One configuration becomes increasingly stable. Others lose redistribution pathways. Threshold approaches.

At articulation, alternatives exhaust. That exhaustion is what we call collapse.

Collapse as Exhaustion

Collapse is not the creation of a state. It is the elimination of alternatives. The state that "remains" was not conjured into being. It was already viable within relational geometry. Measurement did not create it—it constrained competition until only one configuration could survive.

This is articulation exhaustion.

A Mechanical Analogy

Consider a thin sheet under distributed tension. Multiple micro-buckling patterns are possible. As external constraints change, strain redistributes unevenly. At some critical configuration, one buckling pattern becomes dominant. Others cannot sustain geometry under constraint.

The sheet snaps into a specific configuration. From outside, it appears as a sudden jump. From inside the geometry, it was progressive exhaustion.

Quantum collapse follows the same logic.

Continuity Preserved

Before measurement, multiple configurations coexist because relational geometry permits them within finite alignment span. Measurement narrows geometry. One configuration exhausts alternatives. Articulation occurs.

The appearance of discontinuity arises because threshold crossing compresses the final stage of negotiation into a sharp event. But continuity never breaks. Structured continuity remains intact. Only configuration changes.

This reframing dissolves the need for ontological leaps. There is no magical reduction of possibility into reality. There is relational competition resolving under constraint.

Collapse belongs entirely to the relational layer. Ontology remains continuous. Epistemic description updates after articulation.

The drama lies in our interpretation, not in structure.

Collapse as Universal

And once collapse is seen as articulation exhaustion, a powerful insight emerges: the same phenomenon occurs everywhere. In fracture. In chemical reactions. In cognitive decisions. In phase transitions.

Quantum collapse is not special in kind. It is special in scale and visibility.

But before we extend this further, we must clarify what actually changes during collapse. Is it a particle that jumps? Or is it the field configuration that reconfigures?

We now turn to that distinction.

7.2 — Field Reconfiguration, Not Particle Teleportation

If collapse is articulation exhaustion, the next question becomes unavoidable: *What actually changes?*

In popular language, we often say, *"The particle jumps."* It was delocalized, now it is here. It was a wave, now it is a point. But this language already smuggles in a misconception.

It assumes that reality is fundamentally made of isolated particles moving in empty space, and that collapse is some kind of teleportation of a tiny object.

Ontological Reminder

Recall the ontological layer: there are no isolated particles in absolute void. There are configurations within structured continuity.

What we call a *particle* is a stabilized strain geometry—a localized configuration within the field of continuity.

Before measurement, the configuration is not a ghostly object smeared across space. It is a relational geometry whose viable articulations span multiple constraint-consistent configurations.

After measurement, that relational geometry reorganizes under new constraints.

Nothing teleports. *Geometry reconfigures.*

A Mechanical Analogy

Imagine a stretched membrane under tension. Before constraint is applied, several deformation modes are viable. When a new boundary constraint is introduced, the membrane reorganizes. One deformation pattern stabilizes; others lose viability.

Did something jump across the membrane? No. The geometry reconfigured under constraint.

Similarly, when measurement occurs in a quantum system, the measuring apparatus introduces coupling constraints. The relational field—structured continuity under strain—reorganizes.

What changes is not a tiny object deciding where to be. What changes is the *constraint geometry*. The viable configuration that survives threshold becomes the articulated state.

Layered Perspective

- From the **epistemic layer**, this looks like the wavefunction collapsing to a localized value.
- From the **relational layer**, it is constraint-driven exhaustion.
- From the **ontological layer**, continuity persists.

This reframing removes a major source of confusion. The particle is not a bead jumping between positions. It is a localized configuration within a larger structured field.

Collapse does not relocate substance. It reorganizes relational geometry.

Why "Field Reconfiguration" Is Precise

This is why the language of *field reconfiguration* is more precise than *particle collapse.* It aligns with the ontological commitment to continuity.

Even in quantum field theory, particles are excitations of underlying fields. Collapse, in this sense, is not the destruction of a wave and creation of a particle—it is the stabilization of a specific excitation configuration under constraint.

The field does not tear. It reshapes.

Continuity Preserved

This also clarifies why collapse does not violate continuity. There is no ontological discontinuity. There is only geometric reorganization under threshold crossing.

The dramatic language of collapse persists because articulation is sharp relative to measurement bandwidth. The system appears to snap into a state. But what snaps is constraint geometry, not existence itself.

Two Confusions Dissolved

When we reframe collapse as field reconfiguration, two major confusions dissolve:

1. The illusion of particle teleportation.
2. The assumption that reality jumps discontinuously.

Reality does not jump. Constraint geometry shifts. And this shift occurs through interaction.

Toward Observer Coupling

Which brings us to the role of the observer. If collapse is reconfiguration under constraint, then what is the observer actually doing?

We now turn to *observer coupling.*

7.3 — Observer Coupling as Constraint Interaction

If collapse is not a mystical jump, and not a particle teleporting across space, then the most controversial question must now be addressed directly: *What is the role of the observer?*

For decades, the language surrounding quantum mechanics has flirted with a dangerous idea—that consciousness creates reality. That the observer, by looking, forces the universe to choose.

But this confusion arises from **layer mixing.**

- At the **ontological layer**, structured continuity exists independently of awareness.
- At the **relational layer**, alignment propagates under interaction.
- At the **epistemic layer**, knowledge updates when articulation occurs.

Observation is not magic. It is **coupling.**

Measurement as Interaction

When we say *"an observer measures a system,"* what physically happens is interaction between two structured systems—the measured configuration and the measuring apparatus (detectors, screens, sensors, or biological systems).

That interaction introduces new constraints. Constraint geometry changes. Strain redistributes differently under the new coupling. Viability shifts. Collapse follows exhaustion.

No consciousness is required for this structural sequence.

A photon interacting with a detector plate does not wait for awareness. Coupling itself is sufficient to alter relational geometry. The apparatus becomes entangled with the system. Alignment propagates through the combined configuration. Threshold crossing articulates one outcome.

The observer does not create reality. The observer participates in it.

More precisely: *the observer is part of the constraint geometry.*

Dissolving the Mysticism

Before measurement, competing strain geometries coexist because relational constraints permit them. When a measurement device interacts, the system is no longer isolated. It becomes part of a larger coupled structure. The relational landscape changes.

Coupling modifies viability. Collapse is not caused by awareness—it is caused by constraint.

This dissolves the mysticism that often surrounds quantum measurement. Consciousness is not a trigger. Interaction is.

The confusion persists because human observation involves both physical coupling and epistemic awareness simultaneously. When we look at a detector reading, physical interaction has already occurred.

But our conscious awareness updates at the same moment, giving the illusion that awareness caused articulation.

It did not. Coupling did.

Even in experiments where detectors record outcomes without a human present, collapse occurs. Data stored in memory devices show articulated results long before anyone inspects them. The relational event happened during interaction. Awareness merely registers articulation.

Structural Coherence Restored

Observer coupling is:

- Physical interaction.
- Constraint introduction.
- Relational reconfiguration.
- Threshold exhaustion.

Nothing more. No metaphysical intervention. No collapse triggered by thought.

This does not diminish the role of observers. It clarifies it. Observers are systems embedded within structured continuity. When they interact, they alter constraint geometry. That alteration influences which strain geometries remain viable.

But they do not override structure. They participate in it.

Why Measurement Feels Final

And once observer coupling is understood as constraint interaction, the decisiveness of measurement becomes easier to explain. Measurement feels final. Why?

Because articulation under coupling exhausts alternatives simultaneously.

We now turn to why measurement feels so sharp, so conclusive—almost violent in its decisiveness. That sharpness is not mystical. It is geometric.

7.4 — Why Measurement Feels Decisive

Even after reframing collapse as articulation exhaustion, something still feels dramatic.

Measurement feels final.

Before measurement, there are possibilities. After measurement, there is a fact. The transition appears sharp—almost surgical. It feels as though reality snaps into place.

Threshold Compression and Bandwidth Limitation

The answer lies in *threshold compression* and *bandwidth limitation.*

Before measurement, competing strain geometries coexist within finite alignment span. The relational system is negotiating. Redistribution pathways are open. Viability is distributed.

But this negotiation is typically invisible. Why? Because it unfolds below perceptual and instrumental resolution.

Measurement introduces strong constraint coupling. Constraint narrows viability space rapidly. Redistribution pathways shrink. Competing geometries lose sustainability nearly simultaneously as threshold approaches.

At the moment of articulation, alternatives exhaust together. The result appears decisive.

The decisiveness does not arise because reality jumps. It arises because threshold crossing is sharp relative to our bandwidth.

Analogies of Sharpness

Imagine a stretched elastic band gradually increasing in tension. The internal redistribution happens continuously. But when the band snaps, the articulation is abrupt. The gradual buildup is hidden inside the sharp event.

Similarly, in phase transitions—water freezing, magnets aligning—microscopic fluctuations accumulate. At critical temperature, configuration reorganizes quickly. The macroscopic change appears sudden.

Quantum measurement operates under similar geometry. The system plus apparatus evolves relationally. Competing configurations entangle. Viability becomes increasingly skewed toward one configuration. When threshold is reached, articulation occurs.

Because the relational exhaustion happens near-simultaneously across the constrained system, the epistemic update is abrupt. We register only the articulated state. The negotiation vanishes from view.

Irreversibility at Scale

There is another reason measurement feels decisive: **irreversibility.**

Once articulation occurs under coupling, the constraint geometry of the combined system changes. The apparatus now encodes the outcome. Strain redistributes across a much larger network—detector states, environmental degrees of freedom, recording devices.

Reversing that redistribution would require reconstructing prior relational geometry across countless degrees of freedom. Improbable.

Thus measurement appears final because it modifies geometry irreversibly at scale.

Layered View of Sharpness

It is not that collapse creates reality. It is that articulation exhausts alternatives under strong coupling and propagates irreversibly through a larger system.

- From the **epistemic layer**, we see a sudden update.
- From the **relational layer**, we see threshold crossing.
- From the **ontological layer**, continuity remains intact.

The sharpness of measurement is not metaphysical. It is geometric compression.

Collapse as Universal Phenomenon

Once this is understood, collapse loses its mystique. It becomes one instance of a universal structural phenomenon: articulation under constraint.

But to fully dissolve the quantum-classical divide, we must show that this phenomenon is not unique to microscopic systems. Collapse happens everywhere.

We now expand collapse beyond quantum mechanics.

7.5 — Collapse Across Scales

If collapse is articulation exhaustion under constraint, then it cannot be unique to quantum systems. It must be universal. And once we look carefully, we see it everywhere.

Phase Transitions

Water cools gradually. Molecular motion redistributes. Competing configurations coexist—liquid-like fluctuations, solid-like clusters.

As temperature crosses a critical threshold, one configuration exhausts alternatives. Crystalline geometry stabilizes.

We say, *"Water freezes."* But freezing is collapse at scale. A configuration survives threshold exhaustion.

Chemical Reactions

Reactants approach activation energy. Bond strain accumulates. Multiple intermediate configurations compete. At critical energy, one reaction pathway exhausts alternatives. New molecular geometry stabilizes.

We say, *"The reaction occurred."* That too is collapse.

Mechanical Fracture

Strain distributes across material. Micro-cracks form and vanish. Competing crack geometries negotiate. At threshold, one crack path exhausts alternatives. Fracture propagates.

We say, *"It broke."* Collapse again.

Cognition

You deliberate. Competing narratives coexist. Emotional strain and reasoning pathways negotiate. Under sufficient internal constraint, one narrative exhausts alternatives. You decide.

We say, *"I made a choice."* Collapse once more.

The Universal Pattern

In each case:

- Multiple viable configurations exist.

- Constraint reshapes viability.
- Threshold is approached.
- One configuration exhausts alternatives.
- Articulation occurs.

The pattern is identical.

Quantum collapse is not metaphysically special. It is collapse at small scale. The only reason it appears more mysterious is because microscopic alignment spans are comparable to measurement interaction scales. Uncertainty remains visible. Negotiation cannot be averaged out through aggregation.

At macroscopic scales, collapse events aggregate. Redistribution pathways multiply. Micro-articulations blend into smooth classical behavior. But the underlying mechanism is unchanged.

Dissolving the Divide

This realization dissolves the quantum-classical divide. There is not one type of collapse in quantum mechanics and another in classical systems. There is only articulation exhaustion under constraint.

The difference lies in:

- Scale of interaction.
- Density of competing geometries.
- Bandwidth of observation.
- Degree of aggregation.

Quantum collapse is visible because alignment spans are short and constraint coupling is precise. Classical collapse is familiar because we experience it daily—but we do not call it collapse. We call it freezing, reacting, breaking, deciding.

Collapse is universal. Quantum mechanics did not invent collapse. It revealed it at small scale.

Toward Determinism

Once collapse is reframed as articulation exhaustion, the dramatic mystery fades. What remains is geometry.

And now a new insight emerges: if collapse occurs everywhere, and if collapse aggregates at large scale, then classical determinism may not oppose quantum uncertainty. It may emerge from repeated structured collapse events.

The next step is to examine how aggregated articulation produces predictability.

We now move toward the architecture of determinism.

From Collapse to Classical Order

We began this chapter with a dramatic word: *collapse.*

By now, the drama has softened. Collapse is not a jump in ontology. It is not a mystical act of consciousness. It is not randomness erupting into reality.

Collapse is articulation exhaustion under constraint.

Competing strain geometries coexist. Coupling reshapes viability. Threshold is crossed. Alternatives exhaust. Configuration stabilizes.

This process is not unique to quantum systems. It happens in freezing water, in breaking materials, in chemical bonding, in neural decisions, in social revolutions. Quantum collapse is simply the smallest visible instance of a universal structural event.

Dissolving the Divide

Once we see this, the tension between quantum indeterminacy and classical determinism begins to dissolve.

Determinism is not the absence of collapse. It is the aggregation of collapse.

At microscopic scales, articulation events remain visible as probabilistic outcomes because alignment spans are comparable to measurement scales. Uncertainty is exposed.

At macroscopic scales, countless collapse events accumulate and average. Competing geometries resolve continuously across vast relational networks. The aggregated behavior appears smooth, stable, predictable.

Classical order emerges from structured collapse repetition.

A Continuous Structure

This is a profound shift. The universe is not divided into two realms—one probabilistic and one deterministic.

There is one continuous structure.

At small scale, we see negotiation. At large scale, we see aggregated resolution.

Uncertainty does not vanish. It integrates. Collapse does not contradict stability. It builds it.

Toward Determinism

Now the next question becomes clear:

If classical predictability emerges from aggregated articulation, how exactly does scale transform uncertainty into determinism? How does micro-indeterminacy dilute into macro-regularity?

The next chapter addresses this directly.

We now move from collapse itself to how collapse organizes into classical law— from articulation events to the architecture of determinism.

Interlude — What Has Just Changed

Pause for a moment. Something important has shifted.

Uncertainty is no longer mysterious. Collapse is no longer dramatic. Time is no longer a container. Probability is no longer ontological.

This part did not attack quantum mechanics. It clarified it.

Core Realizations from This Part

1. Reality Operates in Layers

- Ontological Layer → Structured continuity (SVE), strain geometry.
- Relational Layer → Alignment, competition, threshold (ATT).
- Epistemic Layer → Probability, equations, measurement models. Confusion arises when these layers are mixed.

2. Ontology Is Continuous

- No isolated particles in empty void.
- No ontological randomness.
- No metaphysical jumps.
- Reality consists of structured configurations under strain. Randomness does not belong to ontology.

3. Relational Dynamics Produce Uncertainty

- Competing strain geometries coexist.
- Alignment requires finite span (ATT).
- Uncertainty lives in pre-threshold negotiation.
- Collapse belongs to relational dynamics, not ontology. Uncertainty is structural, not mystical.

4. Probability Is Descriptive, Not Fundamental

- Probability reflects viability distribution.
- The wavefunction encodes unresolved relational geometry.
- Collapse updates epistemic description after articulation. Probability describes unresolved structure. It does not create it.

5. Resolution Is Never Instantaneous

- Articulation appears sudden.
- Negotiation precedes threshold crossing.
- Instantaneity is perceptual compression.
- Every collapse has geometric preparation. Nothing truly "jumps."

6. ATT Is Structural Duration

- Alignment Threshold Time measures resolution span.
- Delay is not weakness.
- Delay is redistribution capacity.
- Stability stretches delay; fragility compresses it. Time emerges from finite alignment.

7. Time Is Perceived Alignment Span

- Time is not an absolute container.

- Duration reflects alignment unfolding.
- Sequence arises from negotiation.
- The arrow of time follows threshold exhaustion. Without delay, there is no time.

8. Irreversibility Is Geometric

- Threshold changes constraint geometry.
- Alternatives lose viability.
- Reversal requires reconstructing prior strain distribution.
- Entropy reflects distributed strain expansion. The arrow of time is structural, not mystical.

9. Collapse Is Articulation Exhaustion

- Collapse is not creation.
- Collapse is elimination of competing geometries.
- Field reconfiguration replaces particle teleportation.
- Ontology remains continuous. Collapse belongs to the relational layer.

10. Observer Is Coupling, Not Consciousness

- Measurement = constraint interaction.
- Coupling reshapes viability.
- Collapse follows exhaustion.
- Awareness registers articulation; it does not cause it. No mystical observer needed.

11. Measurement Feels Decisive Because Threshold Is Sharp

- Negotiation is hidden below bandwidth.
- Threshold crossing compresses process.

- Irreversibility amplifies finality.
- Epistemic update appears abrupt. Decisiveness is geometric compression.

12. Collapse Is Universal

- Freezing is collapse.
- Fracture is collapse.
- Chemical reaction is collapse.
- Decision-making is collapse. Quantum collapse is not unique. It is small-scale articulation.

13. Determinism Is Aggregated Collapse

- Micro-collapse events accumulate.
- Aggregation smooths indeterminacy.
- Classical laws emerge from structured repetition.
- Uncertainty does not vanish—it integrates. The quantum-classical divide dissolves.

The Core Shift

Quantum mechanics is not mysterious because reality is irrational. It appears mysterious because:

- We mixed layers.
- We mistook description for ontology.
- We mistook relational delay for randomness.
- We mistook articulation for jump.

Once separated:

- Continuity remains intact.
- Collapse becomes structural.

- Time becomes alignment span.
- Determinism becomes aggregated exhaustion.

Part III — Quantum Without Mysticism

"The quantum world is not irrational. It is simply geometry negotiating under minimal scale."

Overview

Quantum mechanics has long been treated as the frontier where logic bends and reality becomes strange:

- Wave–particle duality.
- Superposition.
- Collapse.
- Discreteness.
- Indeterminacy.

These ideas have fueled not only physics, but philosophy, metaphysics, and speculation about consciousness and parallel worlds.

But what if the strangeness is not in reality— but in interpretation?

Structural Foundation

The previous parts established the framework:

- Reality is continuous.
- Strain (*delta*) generates negotiation.

- Alignment requires finite span (ATT).
- Collapse is articulation exhaustion.
- Probability is descriptive, not ontological.

This part now applies that discipline directly to quantum theory.

Reframing Quantum Concepts

This section does not discard Schrödinger, Planck, or modern physics. It reinterprets them structurally:

- **Wavefunctions** reframed as baseline unresolved strain geometry.
- **Superposition** as layered relational competition.
- **Planck's constant** as minimal alignment packet.
- **Discreteness** emerging from threshold behavior under bandwidth limits.
- **Decoherence** distinguished from true resolution.

The Goal

The goal is clarity without mysticism. Quantum mechanics is not a paradox factory. It is a scale-sensitive manifestation of structured delay, minimal bandwidth, and articulation under constraint.

By the end of this part, quantum theory will no longer appear disconnected from classical reality. It will appear as the same structure—observed near minimal alignment span.

We now enter the quantum domain. This time, without metaphysical fog.

Chapter 8 — Wavefunctions as Baseline Delta

What Is Actually Waving?

Few symbols in physics are as iconic as the wavefunction: ψ. We draw it as a smooth oscillation. We square it to obtain probability. We say it spreads, interferes, collapses.

But pause. What exactly is waving?

Is matter literally oscillating across space? Is probability rippling like water? Is reality itself vibrating between possibilities? Or are we mistaking representation for substance?

Schrödinger's Intuition

When Schrödinger introduced his equation, he was not trying to create mysticism. He was searching for continuity. Classical mechanics described trajectories—clean, deterministic paths. But quantum experiments revealed interference patterns that demanded a wave-like description.

So he wrote a wave equation. Not because he believed particles were ghosts, but because continuity demanded a dynamic, field-like representation.

The mathematics worked. But interpretation drifted.

We began to speak as if the wavefunction were a physical mist spread through space. We imagined electrons as clouds of probability. We described particles as existing in multiple places at once. The language became metaphysical.

Returning to Structural Discipline

Yet if we return to structural discipline, a different question emerges: perhaps nothing is waving in the way we imagine.

The wavefunction may not describe a substance at all. It may encode something more fundamental: *unresolved strain.*

Before collapse, before articulation, before threshold exhaustion, competing strain geometries coexist within structured continuity. That coexistence must be represented somehow.

The wavefunction may be that representation.

Not a wave of matter. Not a wave of probability. But a mathematical encoding of *baseline delta*—unresolved geometric distortion distributed across constraint space.

Oscillation as Strain Propagation

The oscillation in ψ may not represent physical rippling of substance. It may represent the dynamic propagation of strain under relational evolution.

If so, then the mystery dissolves.

Wave–particle duality becomes less about contradiction and more about scale-sensitive articulation.

- The "wave" describes unresolved geometry.
- The "particle" describes articulated configuration.

Nothing mystical transitions between them. Only threshold exhaustion.

Reframing the Wavefunction

In this chapter, we reinterpret the wavefunction not as a ghostly physical entity, but as a *structural map of unresolved relational strain.*

- Superposition becomes layered geometry.
- Probability becomes viability density.
- Indeterminacy becomes structural inevitability at minimal alignment span.

We do not discard Schrödinger. We clarify him.
We begin by returning to his original structural intuition.

8.1 — Schrödinger's Structural Intuition

Before the wavefunction became a symbol of paradox, it was an attempt at continuity.

Erwin Schrödinger was not trying to mystify physics. He was responding to a tension. Classical mechanics described particles as discrete objects moving along precise trajectories. But experiments—especially interference phenomena—refused to obey that picture.

Light produced interference. Electrons produced interference. Matter behaved in ways that suggested wave-like structure.

Schrödinger's response was not to abandon structure. It was to restore it.

He wrote a wave equation. Not because he believed particles were literally water waves, but because he sought a continuous dynamical description capable of producing observed patterns.

Determinism at the Core

The Schrödinger equation is not probabilistic in its core form. It is deterministic. Given an initial wavefunction and a Hamiltonian, its evolution is fully specified.

This is critical. The equation evolves ψ continuously. It does not randomly select outcomes. Randomness enters later, through interpretation—specifically, through Max Born's probabilistic reading of $|\psi|^2$.

Schrödinger's original intuition was structural: there must be a continuous dynamical entity whose evolution explains interference and quantization. He was searching for geometry, not mysticism.

Layered Reinterpretation

In our layered framework, that geometry corresponds to unresolved relational strain.

- The wave equation governs how strain propagates under constraints.
- The Hamiltonian encodes constraint geometry.
- The differential operator encodes relational evolution.
- The wavefunction encodes distributed configuration under strain.

Nothing in Schrödinger's equation requires ontological randomness. It describes how a structured configuration evolves in time under constraint.

The later probabilistic interpretation—while extraordinarily successful—shifted attention from structural evolution to outcome distribution. Probability became central. Superposition became mysterious.

But if we separate layers carefully, we can see that Schrödinger's equation naturally fits within the *relational layer.* It describes the evolution of unresolved geometry prior to articulation.

ψ is not necessarily a substance. It is a dynamic configuration descriptor.

Interference Without Contradiction

Interference patterns arise not because particles are schizophrenic, but because relational geometry evolves continuously and can superpose strain contributions across paths.

This is structurally consistent.

Consider the double-slit experiment. The interference pattern emerges from the evolution of ψ across both slits. But nothing in this requires the electron to be in two contradictory ontological states simultaneously. What it requires is that relational strain geometry propagate across both constraint pathways prior to articulation.

The wavefunction encodes that propagation. Collapse later selects an articulation under coupling.

Structural Continuity Preserved

Schrödinger's equation therefore captures the continuous relational unfolding of strain geometry. Born's rule captures the epistemic distribution of articulation outcomes. Collapse captures relational threshold exhaustion.

Seen this way, Schrödinger's intuition aligns perfectly with structured continuity. He preserved continuity at minimal scale.

The mystery arose when representation was mistaken for ontology. ψ was never required to be a ghostly mist. It was required to encode evolving structure.

And once we restore that clarity, the wavefunction becomes less mysterious. It becomes a baseline descriptor of unresolved delta.

We now turn to *probability itself—not as randomness, but as a measure of geometric viability within that unresolved structure.*

8.2 — Probability as Unresolved Geometry

Once the wavefunction is understood as a dynamic descriptor of evolving configuration, the next question becomes inevitable: *What does probability actually represent?*

In standard quantum mechanics, Max Born proposed that the square modulus of the wavefunction, $|\psi|^2$, gives the probability density of finding a particle at a particular location. This interpretation has been experimentally confirmed countless times. It works with extraordinary precision.

But what does that probability mean? Does it mean reality itself is random? Does it mean the particle "chooses" a location arbitrarily? Does it imply indeterminacy is ontological?

Not necessarily.

Probability in the Layered Framework

Within the layered discipline we have established, probability belongs to the *epistemic layer.* It describes outcome distribution under unresolved relational dynamics.

Before measurement, multiple strain geometries coexist within finite alignment span. These geometries correspond to different viable articulation pathways. The wavefunction evolves deterministically according to Schrödinger's equation, encoding how these competing configurations propagate under constraint.

At this stage, nothing random has occurred. There is structured evolution of relational geometry.

However, when coupling occurs—when measurement introduces constraint—the system must articulate one configuration. Because multiple viable geometries exist, and because relational competition has not yet been exhausted prior to coupling, the outcome distribution follows the viability weights encoded in ψ.

$|\psi|^2$ is not a mystical probability mist. It is a *viability density.*

It reflects how strongly a particular configuration is represented within unresolved relational geometry.

Probability as Viability

Probability, in this structural view, measures the relative stability of articulation pathways under given constraint geometry.

- If one configuration carries higher strain coherence under propagation, it will have higher viability density.
- If another carries less, it will appear less frequently across repeated trials.

Repeated experiments do not reveal ontological randomness. They reveal consistent geometric weighting under repeated relational coupling. The statistical distribution emerges from structured competition.

Analogies Beyond Quantum Systems

This is not unique to quantum systems.

- **Mechanical bifurcation:** A system poised at a bifurcation point settles into one of several stable configurations. Small perturbations determine which outcome occurs. Repeated trials produce distributions that reflect geometry, not mystical randomness.

- **Thermal systems:** Molecules follow probabilistic distributions not because reality is random in essence, but because enormous relational degrees of freedom produce outcome distributions under constrained energy landscapes.

Probability is a measure of relational ignorance at the epistemic layer—but structural inevitability at the relational layer.

Quantum Specifics

In quantum systems, because alignment spans are comparable to interaction scales, unresolved competition remains visible. Probability becomes essential to describe outcome distributions.

But that does not elevate probability to ontology. It describes unresolved geometry prior to articulation.

This distinction is subtle but decisive. We are not saying quantum outcomes are predetermined in a classical hidden-variable sense. We are saying they emerge from relational exhaustion of competing strain geometries under constraint.

The distribution of outcomes reflects the geometry of unresolved delta encoded in ψ.

Probability does not inject randomness into structure. It encodes the relative weighting of articulation pathways before threshold exhaustion.

Layered Summary

Seen this way:

- Schrödinger's equation governs structural evolution.

- $|\psi|^2$ encodes viability density.
- Collapse selects an articulation after competition exhausts.

No dice are thrown. Geometry negotiates.

And when negotiation ends under coupling, articulation frequency reflects the weighted structure that existed prior to threshold.

Toward Superposition

With this clarified, we now turn to *superposition*—often the most misunderstood concept in quantum theory.

If probability reflects unresolved geometry, then superposition must reflect layered strain within that geometry.

We now examine superposition without contradiction.

8.3 — Superposition as Layered Strain

Few ideas in quantum mechanics provoke as much discomfort as superposition.

A particle is said to be in two places at once. A spin is up and down simultaneously. A system exists in mutually exclusive states—until measured. The language feels contradictory.

But contradiction emerges only if we mistake layers.

If we treat superposition as ontological duplication, paradox is unavoidable. If instead we treat it as relational coexistence of layered strain geometries, coherence returns.

What Superposition Represents

The wavefunction can be expressed as a linear combination of basis states:

$\psi = a1\phi1 + a2\phi2 + \ldots$

Each term corresponds to a possible articulation under given constraints. These are not separate realities stacked on top of each other. They are relationally viable configurations encoded within unresolved geometry.

Superposition is not a particle literally occupying contradictory positions. It is *layered strain geometry.*

Multiple alignment pathways coexist because finite alignment span has not yet exhausted competition. Constraint geometry permits several viable configurations to remain stable within relational negotiation.

- At the **ontological layer**, structured continuity holds strain.
- At the **relational layer**, strain propagates across multiple constraint-consistent pathways.
- At the **epistemic layer**, we represent that coexistence as superposition.

Nothing ontologically contradictory is required.

Mechanical Analogy

Imagine a structure under load that could buckle along several modes. Before collapse, multiple deformation patterns are mathematically viable. The structure carries strain distributed across these modes.

It is not physically buckled in all modes at once. It is carrying layered strain potential.

When threshold is reached, one mode exhausts alternatives and articulates.

Superposition is similar. The system is not fully articulated in any basis state. It carries layered delta across possible articulation geometries. These layers interfere because strain contributions combine relationally under propagation.

Double-Slit Illustration

Interference patterns in the double-slit experiment illustrate this beautifully.

If an electron were a classical particle choosing one slit, interference would not appear. But if relational strain geometry propagates across both constraint pathways, interference naturally emerges.

The wavefunction encodes that distributed geometry. Superposition reflects that relational distribution.

Dissolving the Paradox

The confusion arises when we ask: *"But where is the particle really?"*

That question assumes ontology at the articulation level before articulation occurs.

Before threshold exhaustion, there is no articulated "particle location" to speak of. There is only layered relational strain across constraint geometry.

Superposition is the structural condition of unresolved competition.

It is not illogical. It is inevitable wherever:

- Multiple articulation pathways are viable.
- Alignment span is finite.
- Constraint geometry permits coexistence.

At quantum scales, these conditions are visible because minimal alignment spans are comparable to interaction scales. At macroscopic scales, layered strain rapidly exhausts into stable configurations, hiding superposition beneath aggregated collapse.

Superposition is therefore not mystical multiplicity. It is *layered delta awaiting articulation.*

Once measurement coupling occurs, constraint geometry narrows. One strain layer exhausts alternatives. Articulation occurs. The layered structure collapses into a single configuration.

The paradox dissolves.

The system was never in contradictory articulated states. It was in unresolved relational geometry.

Toward Indeterminacy

And this leads to the deeper insight: indeterminacy itself is not an anomaly. It is structural at minimal scale.

We now turn to why indeterminacy cannot be eliminated at quantum scales—and why it is inevitable there.

8.4 — Why Indeterminacy Is Structural

At this point, the temptation may arise to think: *"If superposition is layered strain and collapse is articulation exhaustion, then perhaps quantum indeterminacy is only temporary ignorance. Perhaps beneath it, a fully deterministic structure exists."*

This intuition is understandable. But it misses something fundamental.

Indeterminacy at quantum scale is not accidental. It is *structural.*

Conditions for Uncertainty

Uncertainty requires:

- Competing strain geometries,

- Finite alignment span (ATT),
- Limited interaction bandwidth.

At quantum scale, these conditions are not rare—they are dominant.

Minimal scale means:

- Alignment spans are extremely small.
- Interaction timescales are comparable to resolution span.
- Coupling immediately alters constraint geometry.
- Redistribution capacity is minimal.

In such regimes, competition cannot be averaged out.

Scale Contrast

At macroscopic scales, micro-competing geometries collapse continuously across vast relational networks. Aggregation smooths fluctuation. Determinism emerges statistically.

At microscopic scales, aggregation does not yet occur. The system operates near minimal alignment packet. The measurement interaction is not negligible compared to ATT—it is comparable to it.

This is decisive. When measurement timescale ≈ alignment span, unresolved competition remains visible.

Indeterminacy appears not because reality lacks structure, but because structure is negotiating at minimal scale.

Visibility of Competition

The smaller the system, the fewer redistribution pathways exist. Strain geometry cannot diffuse across large networks. Competing configurations remain sharply weighted.

Therefore, indeterminacy is not a failure of determinism. It is the *visibility of relational competition.*

This also explains why attempts to "peek" more precisely into quantum systems often increase indeterminacy. Stronger coupling alters geometry more aggressively. Bandwidth limitations intensify. Alignment span becomes constrained further.

The act of probing changes the relational landscape. Indeterminacy intensifies because geometry becomes more constrained, not because reality becomes irrational.

Minimal Alignment Packet

Another way to see this is through minimal alignment packet behavior.

If there exists a smallest meaningful alignment span—below which articulation cannot be subdivided—then discreteness and indeterminacy are inevitable.

You cannot resolve below the structural bandwidth limit.

At minimal scale:

- Alignment cannot subdivide indefinitely.
- Threshold behavior becomes quantized.
- Competing geometries remain unresolved until coupling.

Thus indeterminacy is built into scale. Not as randomness, but as unavoidable competition under minimal alignment bandwidth.

Beyond Hidden Variables

Hidden-variable approaches often attempt to restore classical determinism beneath quantum indeterminacy. But if relational competition is intrinsic at minimal scale, then determinism in the classical sense cannot exist at that level.

Structure remains deterministic in evolution (Schrödinger's equation), but articulation outcomes depend on relational exhaustion under constraint.

Deterministic evolution + structural competition + finite alignment span = indeterminacy in articulation.

This is not paradox. It is geometry.

Indeterminacy as Structural

Indeterminacy is structural because:

- Alignment is finite.
- Competition is real.
- Measurement alters geometry.
- Aggregation is minimal.

At quantum scale, the system is observed close to its minimal negotiation regime. Uncertainty is not hidden. It is exposed.

The quantum world is not strange because it is irrational. It is strange because we are looking directly at relational negotiation without aggregation.

Toward Synthesis

And now we are ready to synthesize this view:

If the wavefunction encodes baseline delta, if probability measures unresolved geometry, if superposition is layered strain, if indeterminacy is structural—then ψ is not a ghostly wave.

It is a *structural negotiation map.*

We now consolidate that claim clearly.

8.5 — Wavefunction as Baseline Delta Field

We are now in position to state the reinterpretation clearly.

The wavefunction is not a ghostly physical mist. It is not a metaphysical probability cloud. It is not a mysterious dual-state entity.

It is a structured map of baseline delta.

Layered Reframing

- At the **ontological layer**, structured continuity exists under distortion. Strain geometry permeates configuration space. Nothing is isolated. Nothing is random. There is only relational configuration under constraint.
- At the **relational layer**, this strain propagates. Competing articulation pathways coexist within finite alignment span. Redistribution and interference occur across constraint geometry.
- At the **epistemic layer**, we encode this relational state in ψ.

The wavefunction is therefore: *A mathematical representation of unresolved relational strain geometry distributed across possible articulation configurations.*

Its amplitude encodes relative strain coherence. Its phase encodes relational interference structure. Its evolution encodes alignment propagation under constraint.

Nothing in ψ requires ontological duplication of particles. Nothing requires probability to be fundamental substance. ψ is baseline delta expressed mathematically.

Features Explained Naturally

This framing explains several quantum features with structural clarity:

- Interference arises because layered strain geometries combine relationally.
- Superposition reflects coexistence of viable articulation pathways.
- Probability reflects viability density of those pathways.
- Collapse reflects exhaustion of alternatives under coupling.

The wavefunction is not something that exists independently of structure. It is structure described.

More precisely: ψ encodes the geometry of unresolved articulation prior to threshold.

When collapse occurs, the wavefunction does not physically shrink. The relational geometry reconfigures under constraint. The epistemic representation updates accordingly.

Collapse is not a physical tearing of ψ. It is the elimination of layered delta configurations that can no longer sustain under new geometry.

Deterministic Evolution, Indeterminate Articulation

This also clarifies why ψ evolves deterministically under Schrödinger's equation: relational strain propagation is deterministic under given constraints.

What is indeterminate is not evolution—but articulation.

- Evolution follows geometry.
- Articulation follows exhaustion.

Thus we can summarize:

- Wavefunction = baseline unresolved delta.
- Evolution = strain propagation.
- Probability = viability density.
- Collapse = articulation exhaustion.

Nothing mystical remains.

The wavefunction becomes a structural negotiation map—encoding how strain geometry is distributed across potential articulations prior to coupling.

Duality Dissolved

The paradox of wave–particle duality dissolves.

- The "wave" describes unresolved relational strain.
- The "particle" describes articulated configuration after exhaustion.

Both belong to the same structure at different relational stages.

Toward Quantization

With this reframing complete, a deeper question emerges:

If ψ encodes baseline delta, and if indeterminacy emerges at minimal alignment scale, then something fundamental must govern minimal articulation.

Why is energy quantized? Why does discreteness appear? Why does Planck's constant set a lower bound?

The next chapter addresses this directly.

We now move from wavefunction to action. From unresolved strain to minimal alignment packet.

From Baseline Delta to Minimal Action

We began this chapter with a simple question: *What is actually waving?*

We now have a structural answer. Nothing is waving in the mystical sense. What evolves is unresolved strain geometry.

- The wavefunction is not a substance. It is a structural map.
- Superposition is not contradiction. It is layered delta.
- Probability is not ontological randomness. It is viability density of articulation pathways.
- Indeterminacy is not irrationality. It is relational negotiation visible at minimal scale.

With this clarity, quantum mechanics begins to lose its metaphysical tension.

ψ encodes baseline delta. Evolution encodes strain propagation. Collapse encodes exhaustion under coupling.

A Deeper Boundary

But a deeper structural boundary now appears.

If indeterminacy becomes visible at minimal scale, if alignment span cannot shrink indefinitely, if articulation requires threshold—then there must exist a *minimal alignment packet.*

Why does nature exhibit discreteness? Why is energy quantized? Why does Planck's constant appear as a fundamental scale?

Minimal Alignment Bandwidth

If alignment cannot subdivide infinitely, then action itself must have a minimal geometric unit.

Quantum discreteness may not arise from mystical granularity of matter. It may arise from minimal alignment bandwidth.

Planck's constant may not be a mysterious quantum artifact. It may encode the smallest meaningful alignment packet.

Toward Quantization

We now move from wavefunction to action. From layered strain to minimal threshold. From indeterminacy to discreteness.

The next chapter examines *Planck, action, and bandwidth*—and shows how quantization emerges naturally from structural limits.

Quantum mechanics is about to become even less mystical.

Chapter 9 — Planck, Action, and Minimal Bandwidth

Why Nature Refuses Infinity

Classical physics was comfortable with infinity. Energy could be divided endlessly. Time could shrink without bound. Action could approach zero smoothly. Continuity seemed absolute.

But then something subtle disrupted that elegance.

Max Planck introduced a constant—**h.** Energy appeared in packets. Radiation did not vary smoothly at small scales. The product of energy and time refused to fall below a limit.

Why?

Why should nature resist infinite divisibility? Is matter fundamentally granular? Is reality pixelated at its base? Or is there a deeper structural reason why articulation cannot shrink indefinitely?

Structural Principle

Earlier chapters established a foundational principle:

- Resolution requires alignment span.
- Alignment span requires geometry.
- Articulation requires threshold.

If alignment could subdivide infinitely, resolution could occur at arbitrarily small scales without constraint. There would be no minimal event, no smallest meaningful articulation.

But quantum mechanics tells us that such infinite subdivision does not occur.

The Planck Relation

The relation

$$E \cdot T = h$$

suggests something profound.

Energy and time cannot both be arbitrarily small. Their product has a structural floor. This is not merely an experimental quirk. It appears as a fundamental limit.

Perhaps **h** is not a mysterious number inserted into physics. Perhaps it encodes *minimal alignment bandwidth.*

Minimal Alignment Packet

If articulation requires a minimum product of strain intensity (E) and alignment duration (T), then there exists a smallest meaningful event. Below that, no stable articulation can occur.

Continuity remains—but articulation becomes quantized.

Discreteness would not arise from matter being grainy. It would arise from *threshold enforcement in minimal alignment geometry.*

Nature would "refuse infinity" not because it is discontinuous in substance, but because it cannot articulate below minimal bandwidth.

Reframing Planck's Constant

This reframes Planck's constant entirely. Instead of being a magical quantum scale, it may be the *geometric signature of minimal resolution packet.*

The smallest alignment span multiplied by its strain intensity defines the minimal actionable unit. No articulation below that is possible.

- Quantization emerges from threshold.
- Discreteness emerges from minimal alignment packet.
- Infinity is not rejected philosophically. It is limited structurally.

We now begin by reinterpreting the simplest expression:

$$E \cdot T = h$$

What does that product truly encode?

9.1 — Reinterpreting h=E·T

Planck's constant first appeared as a technical fix.

In studying blackbody radiation, Max Planck discovered that classical physics predicted infinite energy at high frequencies—an absurd result known as the *ultraviolet catastrophe.* To resolve it, he introduced the relation:

$$E = h\nu$$

Energy is proportional to frequency. Rewritten,

$$E \cdot T = h$$

since frequency $\nu = 1/T$.

Traditionally, this relation is interpreted as evidence that energy comes in discrete packets—quanta.

But let us slow down and apply structural discipline.

Minimal Articulation Condition

If frequency is inverse time, and if time represents alignment span, then the equation

$$E \cdot T = h$$

may be telling us something deeper. It may be expressing a *minimal articulation condition.*

Earlier, we established:

- Resolution requires alignment span (ATT).
- Strain intensity corresponds to energy density.
- Articulation requires threshold crossing.

Now observe the product E·T.

- Energy measures strain intensity.
- Time measures alignment span.
- Their product measures **action**—the amount of strain integrated over alignment duration.

Planck's constant may therefore represent the smallest possible integrated strain-duration packet required for articulation.

Threshold Enforcement

In other words: no articulation can occur unless the product of strain intensity and alignment span reaches at least h.

Below that, geometry cannot reorganize.

This reframes quantization. Energy is not quantized because matter is granular like sand. Energy exchanges are quantized because articulation requires a minimum integrated strain-duration.

- If strain intensity is small, alignment must persist longer to reach threshold.
- If alignment duration is short, strain intensity must be high to reach threshold.
- Either way, the product cannot fall below h.

Geometry of Delay

This interpretation aligns naturally with the geometry of delay. At minimal scale, alignment span cannot shrink arbitrarily. There exists a floor—below which relational negotiation cannot meaningfully complete.

Thus, h encodes *minimal alignment bandwidth.* It sets the smallest "packet" of actionable relational geometry.

Continuity remains intact. Strain can propagate continuously. But articulation—the act of reconfiguration—cannot occur below minimal action.

Classical vs. Quantum Scales

This also clarifies why classical physics appears continuous at large scales. When action involved in macroscopic processes is vastly greater than h, the minimal packet becomes negligible. Quantization effects disappear beneath aggregation.

But at microscopic scales, where energy and time products approach minimal action, discreteness becomes visible.

Quantization is not ontological fragmentation. It is *threshold enforcement.*

Planck's Discovery Reframed

Planck did not accidentally introduce h. He uncovered the geometric lower bound of articulation—the smallest unit of meaningful structural change.

Once this is seen, discreteness ceases to be mysterious. It becomes inevitable.

But to make this fully clear, we must formalize the idea of a *minimal alignment packet.*

We now turn to that directly.

9.2 — Minimal Alignment Packet

If E·T=h represents a minimal articulation condition, then we must define clearly what this means structurally.

Earlier, we established that articulation requires alignment span. Alignment span (ATT) is the duration over which competing strain geometries negotiate under constraint before threshold exhaustion.

Now we ask: *Can alignment span shrink indefinitely?*

If it could, then articulation could occur with arbitrarily small strain-duration products. There would be no lower bound, no minimal event, no quantization.

But quantum theory shows the opposite. There exists a smallest meaningful action.

This suggests that alignment span has a structural floor. Below a certain duration, relational negotiation cannot complete. Below a certain integrated strain, reconfiguration cannot stabilize.

This minimal articulation unit is what we now call the *minimal alignment packet.*

Defining the Packet

It is not a particle. It is not a chunk of matter. It is the smallest resolvable relational event.

A minimal alignment packet is the smallest product of:

- Strain intensity (energy density), and
- Alignment duration (ATT),

such that threshold exhaustion can occur.

If strain intensity is too weak and duration too short, relational geometry does not reorganize. Competing strain layers persist unresolved. Only when their integrated effect reaches minimal threshold does articulation occur.

Explaining Quantization

This explains why energy exchange happens in discrete quanta.

Consider atomic transitions. An electron does not absorb half a photon because half the minimal alignment packet does not satisfy threshold. The integrated strain-duration required for orbital reconfiguration must reach a discrete structural minimum. Below that, geometry remains stable.

Emission follows the same principle. When strain redistribution occurs within an atom, it cannot radiate arbitrarily small fragments of energy. Articulation requires minimal action packet. Emission therefore occurs in discrete amounts.

Quantization arises from *threshold enforcement at minimal alignment scale.*

Continuity Preserved

Continuity is preserved at the ontological layer. Strain propagates continuously in ψ evolution.

But articulation—the act of structural reconfiguration—requires minimal integrated action.

This is analogous to mechanical systems with activation thresholds. A spring under compression does not release energy until a certain force is reached. Below that, it remains stable.

Similarly, relational geometry at quantum scale requires minimal strain-duration before reconfiguration.

Structural Consequences

The minimal alignment packet defines:

- The smallest meaningful exchange event.
- The smallest resolvable transition.
- The smallest stable articulation.

Planck's constant, in this structural reading, encodes this packet.

It is not the granularity of substance. It is the granularity of articulation.

This distinction is subtle but transformative.

Reality remains continuous. Articulation is discrete.

And once articulation is discrete, energy levels in bounded systems become discrete as well.

We now turn to why discreteness naturally emerges from this minimal alignment condition.

9.3 — Why Discreteness Emerges

If articulation requires a minimal alignment packet, then discreteness is no longer surprising. It becomes inevitable.

Continuous Propagation, Discrete Articulation

Strain can propagate continuously. The wavefunction evolves smoothly. Relational geometry negotiates without jumps.

But articulation—the moment of structural reconfiguration—cannot occur below minimal action. This creates a filter:

- Continuous propagation is allowed.
- Continuous articulation is not.

Bounded Systems

Now consider a bounded system—an atom, for example.

The electron is not a tiny planet orbiting a nucleus. It is a relational configuration constrained by electromagnetic geometry. The wavefunction describes allowed strain distributions under those constraints.

Within that bounded geometry, only certain configurations can satisfy the minimal articulation condition consistently.

Why? Because standing-wave stability emerges from constraint geometry.

If strain must redistribute in a stable manner under boundary conditions, only specific configurations maintain coherence. These correspond to discrete energy levels.

Structural Principles

Discreteness, then, is not imposed from outside. It emerges from two structural principles:

1. Boundary constraint geometry limits viable configurations.
2. Minimal alignment packet enforces threshold condition.

Combine these and the result is quantization.

Only those configurations whose integrated strain-duration product satisfies minimal action *and* boundary coherence can stabilize. All other intermediate configurations fail to meet threshold or fail to maintain coherence under constraint.

This is why energy levels in atoms are discrete. Not because nature is fundamentally pixelated, but because stable articulation under bounded geometry occurs only at specific strain configurations.

Analogy: Vibrating String

Consider a vibrating string fixed at both ends. The string can vibrate continuously in principle. But stable standing waves occur only at specific wavelengths that satisfy boundary conditions.

These discrete modes are not mystical. They are geometric necessities.

Similarly, quantum energy levels correspond to allowed standing strain geometries under minimal alignment constraint.

- Continuous possibilities exist mathematically.
- Stable articulations exist discretely.

Energy Exchange

Now extend this to energy exchange.

When a system transitions from one stable configuration to another, the difference in strain geometry corresponds to a discrete energy difference. That difference must be emitted or absorbed as a minimal alignment packet.

Hence photons appear quantized.

The emission is not chopped matter. It is discrete articulation under threshold enforcement.

Classical vs. Quantum Scales

This also clarifies why classical systems appear continuous. When energy differences involved are vastly larger than minimal action, discrete steps become negligible relative to scale. Quantization is hidden beneath aggregation.

But at microscopic scale, where action approaches h, discreteness becomes visible.

Structural Compression

Thus quantization is not an arbitrary rule inserted into physics. It is structural compression at minimal alignment bandwidth.

Continuous propagation + threshold articulation + bounded geometry = discrete outcomes.

Continuity Preserved

The elegance here is striking. Quantum discreteness is not a violation of continuity. It is continuity constrained by minimal articulation condition.

We now turn to threshold behavior more explicitly—how the minimal alignment packet enforces quantized transitions under dynamic conditions.

9.4 — Threshold Behavior at Minimal Scale

Once the minimal alignment packet is accepted, threshold behavior becomes central.

Quantization is not merely about allowed energy levels. It is about when articulation is permitted to occur.

Threshold governs articulation.

- Below minimal action, no reconfiguration.
- At or above minimal action, reconfiguration becomes possible.

This principle explains why quantum transitions appear sudden and discrete.

Structure of Threshold Behavior

A system under strain can continuously redistribute internal geometry. The wavefunction evolves smoothly. Strain densities oscillate and interfere.

But structural reconfiguration—such as an electron transitioning between energy levels—requires crossing the minimal action threshold.

Until the integrated strain-duration reaches that threshold, nothing changes structurally. Then, once threshold is reached, articulation occurs.

The transition appears instantaneous. But as we have learned earlier, instantaneity is compression of preparation. Strain accumulates. Alignment negotiates. Threshold is approached. Alternatives exhaust.

At quantum scale, because the minimal alignment packet is small, preparation spans are extremely short. The transition therefore appears sudden.

Why Fractional Exchanges Do Not Occur

This threshold behavior also explains why fractional energy exchanges are not observed.

If a system receives energy below the minimal packet, relational geometry does not reorganize. The energy may redistribute elastically or dissipate, but stable reconfiguration does not occur.

Only when minimal integrated action is satisfied does articulation occur.

This explains photon absorption. An atom does not partially absorb a photon because partial absorption does not meet threshold for orbital reconfiguration. The minimal alignment packet must be satisfied.

Similarly, emission occurs only when sufficient strain redistribution allows threshold crossing.

Threshold behavior enforces discreteness dynamically.

Continuous Evolution, Discrete Articulation

Continuous evolution does not imply continuous articulation.

This is the crucial structural insight:

Continuous propagation + threshold enforcement = quantized transitions.

Macroscopic Analogy

We can also see threshold behavior in macroscopic systems.

Consider pushing a block over a ridge. Small pushes produce no motion. Only when force integrated over displacement exceeds frictional threshold does motion occur.

In quantum systems, minimal action plays a role analogous to frictional threshold.

- Below it, no structural shift.
- Above it, reconfiguration.

Threshold and Uncertainty

This also clarifies why the uncertainty principle emerges.

If the minimal alignment packet sets a lower bound on E·T, then precise simultaneous specification of energy and time becomes structurally limited.

It is not epistemic fuzziness. It is bandwidth limitation.

Minimal articulation bandwidth constrains simultaneous localization in complementary variables.

Structural Alignment

Everything aligns:

- Minimal alignment packet defines smallest resolvable event.
- Threshold behavior enforces discrete transitions.
- Quantization emerges naturally.
- Uncertainty reflects bandwidth constraint.

Quantum mechanics ceases to be mystical. It becomes geometry operating at minimal articulation scale.

We now consolidate this perspective by reframing Planck's constant explicitly as structural bandwidth limit.

9.5 — Planck's Constant as Structural Bandwidth Limit

We can now state the reinterpretation without hesitation:

Planck's constant is not the granularity of matter. It is the granularity of articulation.

This distinction changes everything.

For over a century, h has been treated as a mysterious quantum scale—a fundamental constant marking the border between classical and quantum physics. It appears in energy–frequency relations, uncertainty principles, and commutation relations. It sets the scale for quantization.

But what does it mean structurally?

Minimal Bandwidth Condition

In the framework we have built, articulation requires minimal integrated strain-duration. That minimal product is:

$$E \cdot T = h$$

- Energy represents strain intensity.
- Time represents alignment span (ATT).
- Their product represents action—the total strain integrated over alignment duration.

Planck's constant therefore encodes: *The smallest meaningful alignment bandwidth required for structural reconfiguration.*

Below this bandwidth, geometry may propagate—but it cannot reorganize.

Foundational Consequences

This explains several quantum facts simultaneously:

1. **Quantization** — Because articulation requires minimal bandwidth, only discrete transitions occur.
2. **Uncertainty Relations** — Because alignment span cannot shrink below minimal threshold, complementary variables cannot be simultaneously resolved beyond structural limit.
3. **Quantum–Classical Boundary** — Because when action ≫ h, the minimal packet becomes negligible, and classical continuity dominates.

Planck's constant does not fragment reality. It limits resolution.

It defines the smallest "bit" of geometric reconfiguration possible in structured continuity.

Bandwidth Analogy

Seen this way, h is not mystical. It is a bandwidth parameter.

In communication systems, bandwidth limits how much information can be transmitted per unit time. Below a certain signal threshold, nothing meaningful is resolved.

Similarly, at quantum scale, relational geometry cannot transmit or resolve articulation below minimal action bandwidth.

Alignment with Structural Framework

This interpretation aligns beautifully with earlier chapters:

- Collapse is articulation exhaustion.
- ATT defines alignment span.
- Minimal alignment packet defines lowest actionable articulation.
- Threshold behavior enforces quantization.

Discreteness does not contradict continuity. Continuity governs propagation. Bandwidth governs articulation.

Quantum mechanics, in this light, is not about particles behaving strangely. It is about geometry operating at minimal bandwidth.

Scale Contrast

Once scale increases and action becomes vastly larger than h, the minimal packet fades into insignificance. Classical determinism appears smooth.

But at minimal scale, bandwidth becomes visible. Quantum behavior is simply structure near its resolution floor.

Toward Decoherence

With this understanding, the final question of this part emerges naturally:

If minimal bandwidth governs articulation, and if quantization arises from threshold, then what role does the environment play?

Does decoherence itself resolve articulation? Or does it merely dilute relational geometry?

Is decoherence equivalent to collapse—or something fundamentally different?

We now move to the final chapter of this part: From minimal bandwidth to environmental competition. From quantization to decoherence versus true resolution.

From Minimal Bandwidth to Environmental Geometry

We began this chapter by asking a deceptively simple question: *Why does nature refuse infinity?*

The answer now stands clearly.

Continuity is not broken. Articulation is bounded.

Planck's constant does not fragment reality into grains of matter. It defines the minimal bandwidth of structural reconfiguration.

- Strain can propagate continuously.
- Wavefunctions can evolve smoothly.
- Geometry can negotiate without interruption.

But articulation—the moment when configuration reorganizes—requires minimal integrated action. Below that threshold, nothing changes structurally.

Quantization is not metaphysical discreteness. It is *threshold enforcement under minimal alignment packet.*

Energy appears discrete because articulation cannot occur continuously at minimal scale. Uncertainty arises because alignment bandwidth cannot shrink arbitrarily.

The quantum world is not irrational. It is operating near its resolution floor.

Toward the Final Structural Question

Now we approach the final structural question of this part.

If minimal bandwidth governs articulation, then interaction must matter deeply.

A system rarely exists in isolation. It is embedded in environment—thermal fluctuations, electromagnetic fields, surrounding particles. Environmental coupling reshapes constraint geometry.

But does this environmental influence itself resolve articulation?

- Is *decoherence* equivalent to collapse?
- Or does decoherence merely redistribute layered strain without exhausting competition?

In other words: does the environment force resolution—or does it only dilute relational coherence?

Why the Distinction Matters

The distinction is crucial. Because if decoherence is not true resolution, then collapse remains a structural event distinct from environmental noise.

The final chapter of this part examines this carefully.

We now move:

- From minimal bandwidth to environmental competition.
- From quantized articulation to decoherence versus true resolution.

The fog around quantum mechanics thins further. Clarity demands one more distinction.

Chapter 10 — Decoherence vs Resolution

Does the Environment Collapse Reality?

Quantum systems do not live in isolation. An electron is never truly alone. An atom is never sealed from interaction. Even in a vacuum, fields fluctuate.

Reality is relational.

When a quantum system interacts with its environment, something changes. Interference patterns fade. Superpositions become practically invisible. Classical behavior appears to emerge.

We call this **decoherence.**

For decades, decoherence has been presented as the solution to the collapse problem. The story is elegant:

- The environment entangles with the system.
- Phase information disperses.
- Interference terms vanish.
- The system "effectively collapses."

But does it?

Collapse vs Decoherence

Has articulation truly occurred? Or has layered strain merely spread beyond our observational bandwidth?

This distinction is not semantic. It is structural.

Earlier, we defined collapse as **articulation exhaustion under threshold.** Competing strain geometries negotiate within finite alignment span (ATT). When threshold is crossed, one configuration exhausts alternatives. Articulation occurs.

But decoherence does something different.

It distributes layered strain across a larger relational network. It entangles the system with environmental degrees of freedom. It suppresses visible interference.

Suppression is not exhaustion. Dilution is not resolution.

A superposition entangled with environment may no longer display interference locally. But globally, the relational geometry still contains layered strain—now spread across system plus environment.

The Crucial Question

So the question becomes unavoidable:

- Does the environment truly resolve uncertainty?
- Or does it merely hide it?

If decoherence is only relational dilution, then collapse remains a distinct structural event. If decoherence equals resolution, then threshold articulation is redundant.

Structural Separation

This chapter separates these two carefully. We will examine:

- Environment as competing geometry.
- The structural difference between ATT and decoherence timescales.
- Why measurement is not merely environmental noise.
- How this distinction can be tested structurally.

Quantum mechanics does not become clearer by merging concepts prematurely. It becomes clearer by separating them precisely.

The fog around collapse has persisted because dilution and resolution were treated as identical. They are not.

Final Distinction

In this chapter, we finish removing mysticism by drawing that line cleanly.

- Decoherence spreads strain.
- Resolution exhausts it.

Only one of them articulates reality.

10.1 — Environment as Competing Geometry

A quantum system is never an island.

Even in carefully isolated laboratory setups, it interacts—however weakly—with surrounding fields, vacuum fluctuations, thermal radiation, electromagnetic noise, or measurement apparatus. Absolute isolation is an abstraction.

What Is the Environment Structurally?

The environment is not background emptiness. It is additional constraint geometry.

Every environmental interaction introduces new relational pathways. When a quantum system couples with its surroundings, its layered strain geometry does not disappear—it becomes entangled. Strain spreads.

Redistribution, Not Exhaustion

Before environmental coupling, a system may carry layered delta across multiple articulation pathways. The wavefunction encodes that distributed strain geometry.

When the environment interacts:

- Additional degrees of freedom are introduced.
- Constraint geometry expands.
- Strain redistributes across system + environment.
- Coherence between layered strain paths becomes harder to track locally.

This is *decoherence*.

But notice what has happened. The layered strain has not been exhausted. It has been redistributed.

Superposition has not vanished ontologically. It has become entangled across a larger relational network.

The interference terms fade in the reduced description of the system because phase information is no longer locally accessible. But globally, within the enlarged system–environment structure, layered strain still exists.

Decoherence is relational diffusion. It is not articulation.

Analogy

Imagine two coherent waves interacting in a small basin. Their interference is clearly visible. Now imagine connecting that basin to a vast ocean. The interference pattern becomes undetectable locally as energy disperses.

But dispersion is not annihilation. The energy remains within the larger system.

Similarly, environmental coupling disperses relational coherence into many degrees of freedom. The local system appears classical because interference cannot be reconstructed easily.

But articulation—the exhaustion of competing strain geometries—has not necessarily occurred.

Environment as Constraint

The system has become embedded within a broader competing geometry. Environment acts as additional constraint structure, not as final judge.

This is crucial.

Decoherence:

- Suppresses visible interference.
- Distributes phase relationships.
- Expands relational complexity.

It does not automatically:

- Enforce threshold exhaustion.
- Select a unique articulated configuration.
- Eliminate alternative geometries globally.

Why This Matters

This is why decoherence explains why macroscopic objects appear classical—but does not, by itself, explain why a single definite outcome occurs.

The environment increases competition. It does not necessarily complete it.

From a structural perspective, environment introduces more competing geometry. It may accelerate strain redistribution, but redistribution is not the same as exhaustion.

- Collapse requires threshold.
- Decoherence introduces dilution.

The two are related but distinct.

Understanding this distinction prevents a conceptual shortcut that has persisted for decades—the idea that environmental noise alone solves the measurement problem.

It does not.

The next step is to examine this difference more precisely by comparing decoherence timescales with Alignment Threshold Time (ATT).

Only then can we see whether dilution equals resolution—or whether resolution requires something more.

10.2 — ATT vs Decoherence

Now we arrive at the structural distinction that clears the fog: Decoherence timescale is not the same as Alignment Threshold Time (ATT).

They describe different processes.

Defining the Two

- **Decoherence time** refers to how quickly interference between components of a superposition becomes practically unobservable due to environmental entanglement. Phase relationships disperse into many environmental degrees of freedom. The reduced density matrix of the system loses off-diagonal coherence terms. *In* *simple terms: interference visibility fades.*

- **Alignment Threshold Time (ATT)** is the duration required for competing strain geometries to exhaust under constraint and produce articulation. *ATT governs structural resolution. Decoherence governs relational dilution.*

These are not identical.

Speed of Decoherence vs Resolution

Decoherence can occur extremely fast—often unimaginably fast for macroscopic systems. A dust particle in air decoheres in fractions of a second. Interference becomes undetectable almost instantly.

But does that mean articulation has structurally exhausted? Not necessarily.

- Decoherence spreads layered strain across system + environment. It makes interference practically irrecoverable. But the total relational geometry still contains multiple branches—now entangled.
- ATT concerns threshold exhaustion within a given constraint geometry. Collapse requires that competing articulation pathways lose viability under threshold enforcement.

Distinct Roles

Decoherence may:

- Increase effective constraint complexity.
- Reduce local interference.
- Make superposition observationally inaccessible.

ATT determines:

- Whether one configuration exhausts alternatives.
- Whether articulation stabilizes uniquely.
- Whether relational negotiation truly ends.

Analogy

Imagine a debate between two positions in a small room. The conversation is intense and visible. Now open the doors and allow the discussion to spread into a massive crowd. Voices disperse. The debate becomes indistinguishable amid noise.

The visibility of the debate fades. But has one position won? No. The competition has diffused.

Similarly, decoherence diffuses layered strain into environment. Resolution requires threshold enforcement that eliminates competing geometry—not merely hides it.

Critical Distinction

There are situations where decoherence occurs without clear articulation. Interference disappears locally, yet the global entangled state still contains multiple coherent branches.

In such cases:

Decoherence time $\ll$ ATT

This is critical. Decoherence suppresses observable interference faster than articulation exhausts relational competition. The environment makes the system appear classical before structural resolution is fully complete.

Why Measurement Is Different

This distinction explains why decoherence solves the *classical appearance problem*—but does not fully solve the *outcome problem.*

Why does a single definite outcome appear in measurement? Because measurement is not mere environmental noise. Measurement is structured threshold enforcement.

Which brings us to the next crucial distinction: What differentiates environmental entanglement from measurement protocols? Why does measurement feel decisive while environmental decoherence merely hides interference?

We now turn to measurement as controlled constraint geometry.

10.3 — Measurement Protocols as Structured Threshold Enforcement

If decoherence is dilution, then measurement must be something more.

Measurement is often described loosely as "interaction with the environment." But this simplification hides a crucial structural distinction.

All measurements involve environmental coupling. Not all environmental couplings are measurements. The difference lies in threshold enforcement.

Measurement as Engineered Constraint

When a quantum system interacts with a measuring apparatus, the apparatus is not just a passive environment. It is engineered constraint geometry.

It is designed to amplify microscopic strain into macroscopic articulation. This amplification is decisive.

Consider a photon striking a photodetector. The interaction does not merely entangle photon and detector at a subtle phase level. It triggers a cascade:

- Electron excitation.
- Charge multiplication.
- Electrical signal amplification.
- Macroscopic readout.

At each stage, strain redistribution crosses thresholds. Competing geometries do not merely diffuse—they are eliminated through amplification.

Amplification as Enforcement

Measurement protocols are designed to push relational geometry across articulation thresholds deliberately.

- Decoherence spreads strain.
- Measurement amplifies it.

Amplification is the key.

When a system interacts with ordinary environment—thermal fluctuations, weak electromagnetic noise—the relational geometry becomes entangled but not necessarily amplified into a stable macroscopic configuration.

But in measurement, the apparatus is constructed to ensure that once the minimal alignment packet is triggered, the process cascades irreversibly.

This cascade forces threshold exhaustion at large scale. It converts microscopic layered strain into macroscopic ripple lock.

Why Measurement Feels Decisive

This is why measurement feels decisive.

The apparatus narrows viability sharply. Constraint geometry becomes highly asymmetric. Once one articulation pathway begins amplifying, it rapidly suppresses alternatives through structural dominance.

Notice how different this is from simple decoherence.

- Decoherence distributes layered strain across environment without necessarily enforcing macroscopic stability in a particular branch.
- Measurement enforces stability by design. It ensures that one articulation pathway becomes self-reinforcing.

Threshold crossing at microscopic scale becomes irreversible macro-configuration.

Persistence of Outcomes

This is also why measurement outcomes persist. A detector reading is stable because ripple lock occurs across vast degrees of freedom. Reversing it would require reconstructing prior strain geometry across countless particles.

Thus measurement = controlled, amplified threshold enforcement. Environment alone ≠ measurement.

Resolving Confusions

This distinction explains several longstanding confusions:

- Why does decoherence explain classical appearance but not unique outcomes? Because decoherence suppresses

interference without necessarily enforcing exclusive articulation.
- Why does measurement produce definite results? Because measurement protocols are built to drive strain beyond threshold and amplify it irreversibly.

Collapse, then, is not caused by mere interaction. It occurs when interaction plus amplification forces relational geometry into irreversible articulation.

Structural Summary

In structural terms:

- Environment adds competing geometry.
- Decoherence diffuses layered strain.
- Measurement enforces threshold through amplification.
- Collapse follows articulation exhaustion under amplified constraint.

Now a final question remains. If this distinction is real, it must be testable.

Can we distinguish decoherence suppression from true structural resolution experimentally?

We now turn to structural testability.

10.4 — Structural Testability

If the distinction between decoherence and resolution is real, it must be testable. Otherwise, it remains philosophical.

The structural claim of this chapter is precise:

- Decoherence suppresses interference visibility through

entanglement.
- Resolution requires threshold exhaustion under constraint.
- Decoherence time and ATT are not identical quantities.
- Measurement protocols enforce amplification; environment alone does not.

These are not vague statements. They generate operational differences.

1. Decoherence Without Resolution

If decoherence merely diffuses layered strain across system + environment, then in principle the global entangled state still contains multiple articulation pathways.

In such regimes:

- Local interference vanishes.
- The reduced density matrix appears classical.
- But global coherence is not structurally exhausted.

This implies that under sufficiently controlled isolation, interference could be partially restored. Indeed, quantum eraser and recoherence experiments already demonstrate that loss of interference does not necessarily mean fundamental collapse. Phase information can sometimes be recovered if environmental coupling is reversed before amplification cascades irreversibly.

Suppression ≠ exhaustion.

2. Comparing Decoherence Time and ATT

If ATT governs articulation exhaustion, then there should exist regimes where:

- Decoherence time is extremely short,
- But articulation threshold is not yet crossed.

In such cases, interference visibility would vanish rapidly, yet the system might still not have undergone irreversible structural reconfiguration.

Conversely, measurement protocols designed to amplify microscopic strain should push the system across threshold rapidly, making reversal practically impossible.

Structural prediction: true resolution correlates with irreversible amplification across macroscopic ripple lock, not merely with phase diffusion.

3. Delayed-Choice and Weak Measurement

Delayed-choice experiments and weak measurements are particularly illuminating.

In weak measurement regimes:

- Coupling is insufficient to drive full threshold exhaustion.
- Information is extracted without enforcing complete articulation.
- Interference patterns can still persist.

This suggests that measurement strength matters structurally. If decoherence alone equaled collapse, then any environmental interaction would fully resolve articulation. But weak measurement shows that partial coupling does not necessarily eliminate relational competition entirely.

Resolution requires sufficient integrated strain-duration (minimal alignment packet) plus amplification.

4. Structural Hierarchy

The framework predicts a clear hierarchy:

1. Pure isolation → Full superposition visible.
2. Weak environmental coupling → Decoherence begins; interference suppressed locally.
3. Controlled measurement without amplification → Partial collapse-like behavior; reversible under some conditions.
4. Amplified measurement protocol → Irreversible threshold exhaustion; stable articulation.

The key variable is not merely entanglement. It is threshold enforcement under minimal alignment packet plus amplification cascade.

5. Why This Matters

For decades, the measurement problem has been treated as metaphysical. But if decoherence and resolution are structurally separable, then collapse is not mysterious.

It is a threshold event.

- Decoherence explains classical appearance.
- Resolution explains definite outcomes.
- One dilutes. The other exhausts.

The distinction is not philosophical decoration. It is operational geometry.

And once collapse is framed as threshold exhaustion under amplified constraint, quantum mechanics integrates seamlessly with earlier chapters:

- Collapse is not mystical.
- It is articulation under bandwidth limit.
- Environment modifies geometry.
- Measurement enforces threshold.

The fog thins again. Now we consolidate this chapter and complete this part.

10.5 — Collapse as Structural Event

We can now state the distinction cleanly, without hesitation:

Decoherence is relational dilution. Resolution is structural exhaustion.

They are not the same.

- Decoherence spreads layered strain across a larger relational network. It suppresses interference visibility by distributing phase information into environmental degrees of freedom. Locally, the system appears classical. But globally, the relational geometry may still contain competing articulation pathways.
- Resolution, however, is stronger. It occurs when competing strain geometries lose viability under threshold enforcement. One configuration stabilizes. Alternatives no longer sustain—even in principle—because constraint geometry plus amplification has eliminated their structural support.

This is collapse.

Collapse is not the fading of interference. It is the elimination of competing geometry.

Irreversibility

This difference becomes evident when we consider irreversibility.

Environmental decoherence can, in principle, be reversed under extraordinary control (as shown in recoherence experiments). But once amplification cascades across macroscopic degrees of freedom—once ripple lock stabilizes across large-scale structure—reversal becomes effectively impossible.

That is threshold exhaustion.

Collapse is not caused by consciousness. It is not caused by mere interaction. It is caused by structural threshold crossing under sufficient integrated action and amplification.

Integration

This integrates all previous chapters seamlessly:

- Wavefunction encodes baseline delta.
- Probability encodes viability density.
- Superposition reflects layered strain.
- Planck's constant defines minimal alignment packet.
- Threshold behavior enforces quantization.
- Decoherence diffuses strain.
- Measurement amplifies strain.
- Collapse exhausts competing geometry.

Nothing mystical remains. Quantum mechanics becomes a special case of universal structural dynamics operating near minimal bandwidth.

At small scale, layered strain is visible. At larger scale, amplification enforces articulation. At macroscopic scale, aggregation produces determinism.

Collapse is not a metaphysical mystery. It is a structural event. Articulation under threshold.

Closing the Quantum Domain

And with this, the quantum domain stands integrated into the broader architecture of uncertainty.

We can now move beyond quantum mechanics. Because once collapse is understood structurally, the same principles must govern scale transitions, classical emergence, extreme strain regimes, and even human cognition.

The next part widens the lens: From quantum bandwidth to macroscopic density. From minimal articulation to persistent reality.

The Quantum Domain Without Fog

We began this part with a promise: *Quantum mechanics, without mysticism.*

We now stand at the end of that promise fulfilled.

- The wavefunction is not a ghostly cloud. It is baseline unresolved delta.
- Probability is not ontological randomness. It is viability density within layered strain.
- Superposition is not contradiction. It is relational coexistence under finite alignment span.
- Planck's constant is not mystical granularity. It is minimal alignment bandwidth.
- Quantization is not fragmentation of matter. It is threshold enforcement under minimal action.
- Decoherence is not collapse. It is dilution of layered strain.
- Measurement is not mere interaction. It is amplified threshold enforcement.

- Collapse is not magic. It is articulation exhaustion.

The quantum world is not irrational. It is structure observed near its resolution floor.

Clearing the Paradox

What appeared paradoxical was not nature—it was interpretation mixing layers.

Once ontological continuity, relational dynamics, and epistemic description are separated cleanly, quantum mechanics becomes elegant again.

Scale and Visibility

But something larger now becomes visible.

If minimal bandwidth governs articulation at small scale, if threshold enforces stability, if amplification produces persistence—then scale must transform visibility.

- At small scale, uncertainty is exposed.
- At large scale, aggregation hides it.

Toward Persistence

The next part widens the lens. We now move:

- From minimal alignment packets to density and persistence.
- From quantum bandwidth to classical appearance.
- From local articulation to large-scale stability.

The world that appears certain is built from the same negotiation we have just examined.

Nothing changes. Only scale does.

Interlude — The Quantum Was Never Magical

Take a breath. The strangest part of physics is now behind us.

- Wave–particle duality.
- Superposition.
- Quantization.
- Collapse.
- Decoherence.

For a century, these ideas have carried an aura of mystery.

But what have we actually discovered?

Not that reality is irrational. Not that consciousness creates matter. Not that the universe plays dice.

What we discovered is **structure operating at minimal bandwidth.**

Core Realizations

1. The Wavefunction Is Baseline Delta

- ψ does not describe a ghostly substance.
- It encodes unresolved relational strain geometry.
- Its evolution is deterministic propagation of strain.
- It maps negotiation prior to articulation. *Wavefunction = structured uncertainty, not mystical wave.*

2. Probability Is Viability Density

- $|\psi|^2$ represents weighting of articulation pathways.
- Probability belongs to the epistemic layer.
- It measures unresolved competition.
- It does not imply ontological randomness. *Probability describes geometry. It does not create it.*

3. Superposition Is Layered Strain

- Multiple articulation pathways coexist structurally.
- Superposition is relational coexistence, not contradiction.
- Interference reflects strain interaction.
- No particle is "in two places" ontologically. *Superposition = layered delta awaiting threshold.*

4. Indeterminacy Is Structural at Minimal Scale

- Alignment span (ATT) is finite.
- Measurement timescale ≈ alignment span at quantum scale.
- Competition remains visible because aggregation is minimal.
- Indeterminacy emerges from limited bandwidth, not irrationality. *Quantum uncertainty is scale-sensitive visibility of negotiation.*

5. Planck's Constant Defines Minimal Alignment Packet

- $E \cdot T = h$ encodes minimal articulation bandwidth.
- Action cannot shrink below structural threshold.
- Energy exchanges are discrete because articulation requires minimal packet.
- Quantization arises from threshold enforcement.

Discreteness is granularity of articulation, not granularity of matter.

6. Threshold Behavior Governs Quantum Transitions

- Continuous propagation does not imply continuous articulation.
- Reconfiguration requires minimal integrated strain-duration.
- Below threshold → no structural shift.
- Above threshold → articulation occurs. *Quantum jumps are threshold events.*

7. Decoherence Is Not Collapse

- Environment introduces competing geometry.
- Decoherence spreads layered strain across system + environment.
- Interference visibility fades.
- But global competition may remain unresolved. *Dilution ≠ exhaustion.*

8. Measurement Enforces Amplified Threshold

- Measurement protocols amplify microscopic strain.
- Amplification cascades enforce ripple lock.
- Competing geometries lose viability irreversibly.
- Collapse follows structural exhaustion. *Collapse = amplified articulation under constraint.*

9. Collapse Is Structural, Not Mystical

- Not caused by consciousness.

- Not caused by randomness.
- Not caused by mere interaction.
- Caused by threshold exhaustion under amplified geometry.
  Quantum mechanics is geometry near minimal bandwidth.

The Core Shift of Part III

The quantum world is not bizarre. It is simply:

- Structure operating near minimal articulation scale.
- Uncertainty visible because aggregation is small.
- Discreteness emerging from bandwidth limits.
- Collapse occurring through threshold enforcement.

What felt mystical was compression of scale.

Once minimal bandwidth is understood, quantum mechanics integrates seamlessly into the larger architecture of uncertainty.

What Comes Next

If minimal bandwidth governs quantum articulation, if amplification produces persistence, if aggregation hides indeterminacy—then scale must be the next frontier.

Why does the classical world appear stable? How does determinism emerge from layered collapse? What happens under extreme strain conditions?

We now widen the lens:

- From minimal packets to large-scale density.
- From quantum articulation to persistent reality.

The universe was never magical. It was negotiating.

Part IV — Scale, Density, and Persistence

"Certainty is not the absence of uncertainty. It is uncertainty aggregated beyond visibility."

Overview

Quantum mechanics revealed uncertainty at minimal bandwidth. But we do not live at minimal bandwidth. We live at scale.

Tables do not flicker between positions. Mountains do not exist in superposition. Human bodies do not decohere into probability clouds.

The classical world appears solid, stable, predictable.

Why?

This part answers without abandoning anything we have built.

Stability is not the elimination of strain. It is strain diluted across density.

Determinism is not the opposite of uncertainty. It is aggregated collapse.

Scale and Persistence

As scale increases:

- Alignment spans overlap.
- Ripple locks reinforce each other.

- Competing geometries exhaust continuously.
- Minimal packets become negligible relative to total action.

The result is persistence.

But persistence is not perfection. Residual strain remains. Extreme density compresses alignment. Black holes become laboratories of overload. Nothing is ever fully resolved.

Themes of This Part

This part explores:

- Why classical reality appears certain.
- How aggregation hides quantum negotiation.
- What extreme strain reveals about collapse.
- Why reality remains dynamically incomplete at all scales.

If the previous part removed mysticism from the quantum world, this part removes illusion from the classical world.

Certainty will be reframed. Determinism will be localized. Persistence will be revealed as tension held across scale.

Transition

We now move from minimal alignment packets to density, aggregation, and endurance.

The world that feels solid is built from continuous negotiation.

Chapter 11 — Why the Classical World Appears Certain

Why Tables Do Not Tremble

If everything we have described so far is true—if uncertainty is structural, if superposition reflects layered strain, if collapse is threshold exhaustion—then a simple question demands honesty:

Why does the world feel solid?

Why does the table in front of you not flicker between positions? Why does your body not dissolve into probability clouds? Why does the sun rise with predictable rhythm?

If quantum mechanics governs all scales, why does the classical world feel certain?

This question is not trivial. It is the heart of the quantum–classical divide.

For decades, explanations have ranged from "quantum effects are too small to matter" to "decoherence makes things classical." But these explanations often stop at description. They do not explain structurally why uncertainty becomes invisible.

Structural Continuity

Nothing in previous chapters suggested that uncertainty disappears at large scale. On the contrary:

- Strain is universal.
- Alignment span is finite everywhere.

- Collapse occurs continuously.
- Minimal bandwidth governs articulation.

So uncertainty must still be present in macroscopic systems. Yet we do not experience it.

Why?

Because visibility depends on scale.

At quantum scale, minimal alignment packet is comparable to the total action involved. Competing geometries are few. Aggregation is minimal. Negotiation remains visible.

At classical scale, systems contain astronomical numbers of interacting degrees of freedom. Alignment events occur continuously and massively. Competing deltas distribute across networks so vast that individual fluctuations become negligible relative to total structure.

Uncertainty does not vanish. It becomes diluted.

Scale Compresses Visibility

Imagine watching a single ripple on a pond. Its motion is visible, delicate, unpredictable. Now imagine the entire ocean. Individual ripples still exist—but they are embedded in collective motion so immense that small fluctuations do not alter the perceived stability of the surface.

Scale compresses visibility.

The classical world feels stable not because it is free from strain, but because strain is distributed across density.

Every atom in the table is negotiating strain. Every molecule is collapsing continuously. Every micro-alignment event occurs under minimal bandwidth.

But these events overlap, average, and reinforce one another.

The result is *persistence*.

Certainty emerges not from the elimination of uncertainty—but from its aggregation.

The table does not tremble because trillions upon trillions of micro-collapses occur every instant, reinforcing large-scale ripple lock.

Stability is dynamic. It is tension held collectively across scale.

This chapter explains how:

- Strain dilution hides micro-negotiation.
- Ripple locks reinforce persistence.
- Aggregated collapse produces determinism.
- Classical certainty emerges locally, not absolutely.

The world does not escape uncertainty. It integrates it.

11.1 — Strain Dilution Across Density

At microscopic scale, strain is visible because it is concentrated.

A single particle carries limited degrees of freedom. A small quantum system negotiates across few articulation pathways. Minimal alignment packets matter directly. Competition is sharp.

But scale changes geometry.

Consider what happens when we move from a single particle to a macroscopic object.

A table contains on the order of 10^{23} atoms. Each atom interacts with neighbors. Each electron carries strain geometry. Each nucleus participates in electromagnetic coupling. Each bond maintains local alignment.

Now imagine the total number of competing strain geometries active at any moment. It is astronomical.

In such density, individual fluctuations do not disappear—they disperse.

This is *strain dilution.*

Structural Transition

Strain dilution means that local delta no longer dominates global geometry. It distributes across vast relational networks. The relative impact of any single articulation event becomes negligible compared to total structural configuration.

At quantum scale:

- Alignment span is minimal.
- Action involved is comparable to minimal packet.
- Competition between strain geometries is limited.
- Collapse events visibly influence system behavior.

At classical scale:

- Total action is enormous compared to h.
- Countless micro-collapses occur continuously.
- Competing geometries overlap massively.
- Local fluctuations average statistically.

The minimal alignment packet becomes insignificant relative to total action.

This is the critical transition.

When total integrated action $\gg$ h, discreteness becomes invisible. Quantization does not vanish—it becomes negligible relative to aggregate scale.

Analogy

Consider a simple analogy.

Imagine flipping one coin. The outcome is uncertain. The probability is evident.

Now flip a billion coins simultaneously. The exact outcome of each coin remains uncertain. But the overall distribution becomes extremely predictable. The average stabilizes. Fluctuations cancel.

The uncertainty did not disappear. It diluted into statistical stability.

Macroscopic objects behave similarly.

Every micro-component negotiates uncertainty under finite alignment span. But because trillions of such negotiations occur simultaneously, and because interactions couple them densely, fluctuations dampen through averaging.

Small deviations cannot destabilize large-scale ripple lock easily. Density absorbs perturbation.

This explains why thermal noise does not cause your table to evaporate spontaneously. Individual micro-events are real. But they are embedded in such dense interaction networks that ripple lock persists.

Strain dilution does not eliminate uncertainty. It reduces its macroscopic visibility.

And this dilution increases with scale.

The larger and denser the system, the smaller the relative influence of minimal alignment packet.

Thus classical stability emerges not from absence of quantum uncertainty, but from overwhelming aggregation of micro-resolution events.

Transition to Reinforcement

However, dilution alone is not sufficient. Density must also reinforce structure.

Which brings us to the next mechanism: *large-scale ripple lock.*

How does aggregation not merely dilute strain, but actively stabilize configuration?

We now turn to persistence through reinforcement.

11.2 — Large-Scale Ripple Lock

Strain dilution explains why microscopic uncertainty becomes less visible at scale. But dilution alone does not produce solidity.

For classical stability to emerge, something stronger must happen: articulations must reinforce one another.

This is where large-scale ripple lock enters.

Ripple Lock at Scale

Recall what ripple lock means structurally. At minimal scale, collapse is articulation exhaustion under constraint—one configuration stabilizes because competing geometries lose viability.

At macroscopic scale, collapse events are not isolated. They occur in dense networks.

- Each micro-articulation modifies local constraint geometry.
- That modification influences neighboring degrees of freedom.
- Neighboring configurations adjust accordingly.
- Alignment propagates.

When trillions of such articulations occur continuously and mutually reinforce one another, a large-scale ripple lock forms.

This ripple lock is not static. It is dynamic equilibrium sustained through continuous micro-resolution.

Examples

Crystal lattice: Each atom sits in a stable configuration relative to its neighbors. Thermal vibrations constantly perturb positions. Strain is continuously negotiated. Yet the lattice persists. Why? Because local articulations reinforce global geometry. If one atom shifts slightly, surrounding bonds redistribute strain and restore alignment. Stability is maintained by collective reinforcement.

River flow: Individual water molecules are in constant motion. Turbulence exists. Micro-uncertainty dominates at the molecular level. Yet the river has shape. It follows a predictable path. Large-scale flow emerges from continuous micro-negotiation reinforced by boundary constraints.

Similarly, macroscopic objects are not rigid because atoms are frozen. They are stable because micro-collapses reinforce collective configuration. Each articulation feeds into a dense web of neighboring articulations. The system becomes self-stabilizing.

Perturbations are absorbed and redistributed. Small fluctuations cannot easily propagate into large-scale instability because ripple lock absorbs them.

Structural Insight

This also explains why classical objects resist small disturbances. Tap a table lightly—it vibrates but returns to equilibrium. Strain redistributes across trillions of bonds. Alignment re-establishes quickly because density supports stability.

Large-scale ripple lock is therefore: Collective reinforcement of articulation across dense relational networks.

It is not absence of strain. It is strain continuously resolved in a mutually stabilizing manner.

This leads directly to aggregated stability.

11.3 — Aggregated Stability

Once strain is diluted and ripple lock reinforces configuration, a new phenomenon emerges: predictability.

Not perfect predictability, but sufficient regularity to form laws. This is aggregated stability.

Structural Averaging

At microscopic scale, collapse events are discrete and visible. At macroscopic scale, collapse events occur continuously and densely. Their cumulative effect smooths fluctuations.

Every atom in a classical object undergoes continuous micro-articulation. Thermal vibrations cause alignment shifts. Electromagnetic interactions fluctuate. Minimal alignment packets are constantly involved.

But because:

- The number of degrees of freedom is enormous,
- The interactions are dense and overlapping,
- The total action involved is vastly greater than h,

individual threshold events do not dominate the system. Instead, micro-collapses average out.

This averaging is not mystical. It is statistical reinforcement under density.

Coin Analogy

One coin flip is uncertain. A million coin flips follow a predictable distribution. The uncertainty remains at the level of each coin. But at scale, fluctuations converge toward stable patterns.

Similarly, microscopic articulation events remain uncertain within structural limits. But when trillions of such events overlap, their aggregate behavior becomes stable and predictable.

Classical Trajectories

This is how classical trajectories emerge.

Consider a thrown ball. At microscopic level, its atoms are vibrating. Electrons are negotiating strain geometry. Minimal alignment packets are constantly active.

Yet the ball follows a smooth parabolic trajectory.

Why? Because the total action governing its motion is astronomically larger than h. The minimal alignment packet is negligible relative to the aggregate strain integrated across the entire object.

Thus, the macroscopic path becomes effectively deterministic.

Determinism here does not mean absence of uncertainty. It means uncertainty is statistically suppressed through aggregation.

The law of large numbers becomes geometry at scale.

Continuous micro-collapses do not create chaos at macroscopic level. They stabilize it.

Classical Laws

This explains why classical laws—Newton's equations, thermodynamics, fluid mechanics—work so reliably.

They describe aggregated articulation patterns. They do not describe elimination of uncertainty. They describe emergent regularity from dense ripple lock.

Conditional Determinism

Importantly, determinism emerges only within bounded regimes.

As long as:

- Total action ≫ minimal packet,
- Density remains high,
- Constraint geometry is stable,

aggregation suppresses visible fluctuation.

But this determinism is conditional. It depends on scale and constraint.

Transition

Which brings us to the next insight: Determinism is local.

It is not an absolute property of reality.

We now examine why classical certainty holds only within specific strain regimes—and how extreme conditions reveal uncertainty again.

11.4 — Why Determinism Is Local

By now, the classical world appears less mysterious.

Uncertainty did not disappear. It diluted across density. Micro-collapses aggregated into stable patterns. Determinism emerged from reinforcement.

But a crucial correction must be made: determinism is not universal. It is local.

Conditional Stability

Determinism holds only within bounded strain regimes—where density, constraint geometry, and total action maintain stable ripple lock.

Within such regimes:

- Minimal alignment packets are negligible relative to total action.
- Fluctuations average out.
- Aggregated stability dominates.
- Trajectories appear smooth and predictable.

But this stability depends on structural conditions. Change the conditions—and determinism weakens.

Fragile Regimes

Chaotic systems: Weather patterns follow deterministic equations, yet small perturbations amplify over time. The system remains deterministic in principle—but local ripple lock is weak. Sensitivity to initial conditions exposes underlying instability.

Phase transitions: A magnet near critical temperature fluctuates wildly. Strain geometry becomes sensitive. Aggregation no longer suppresses micro-variation effectively. Determinism appears fragile near threshold.

Extreme conditions: At high energy densities—inside stars, near black holes, in particle accelerators—minimal alignment packet becomes relevant again. Aggregation no longer dwarfs h by enormous margins. Uncertainty resurfaces visibly.

Locality of Laws

Determinism was never absolute. It was conditional upon scale and density.

This also explains why classical physics fails at certain extremes. Newtonian mechanics works within specific velocity and gravitational regimes. Outside those regimes, relativistic effects emerge. At even smaller scales, quantum effects dominate.

Each layer of determinism is local to a strain regime. No single deterministic framework governs all scales universally.

Determinism is stable ripple lock under specific constraint geometry. When geometry shifts—through extreme energy, density compression, or boundary changes—local stability dissolves.

Human Perception

The illusion of universal certainty arises because we inhabit a strain regime where aggregation is overwhelming and stable. Our sensory systems evolved within this regime. Our cognition assumes continuity and predictability because ripple lock at human scale is extraordinarily dense.

But under structural stress, cracks appear.

Determinism is therefore:

- Emergent from aggregation.
- Stable within density regimes.
- Vulnerable at thresholds.
- Conditional upon geometry.

Classical certainty is real—but not absolute. It is local persistence under bounded strain.

And once we accept this, the final illusion dissolves: solidity is not perfection. It is dynamic tension held across scale.

11.5 — The Illusion of Absolute Solidity

We now arrive at the final refinement of this chapter.

The classical world does not merely appear stable. It appears solid.

The table feels firm. The ground resists your weight. Your body maintains shape. Solidity feels absolute.

But structurally, solidity is not the absence of motion. It is the dominance of reinforcement.

Microscopic Motion

At microscopic scale, atoms in a solid are not static. They vibrate. Electrons redistribute strain continuously. Bonds stretch and relax. Micro-collapses occur constantly.

Nothing inside the table is frozen. Yet the table resists deformation.

Why? Because ripple lock is dense and layered.

Reinforcement Across Density

In a solid, constraint geometry is tightly packed. Alignment spans overlap massively. Every articulation event reinforces neighboring articulations. Strain redistribution is rapidly absorbed across trillions of coupled interactions.

When you press your hand against the table, electromagnetic repulsion between electron clouds prevents interpenetration. That repulsion is not rigid substance—it is structured strain geometry resisting reconfiguration.

The resistance you feel is threshold enforcement across dense alignment networks. Solidity is enforced by collective reinforcement of minimal articulation events.

It is a macroscopic echo of quantum threshold behavior.

Sustained Equilibrium

Because aggregation is enormous, the minimal alignment packet becomes negligible relative to total action. Discreteness vanishes from perception. What remains is stable configuration under continuous micro-resolution.

Thus solidity is not ontological permanence. It is sustained equilibrium. It is tension held gracefully across density.

Conditional Solidity

This also explains why solidity can fail. Under sufficient stress—mechanical force, thermal energy, gravitational compression—ripple lock can break. Bonds snap. Phase transitions occur. Stability dissolves.

Solidity was conditional all along.

The illusion of absolute certainty arises because:

- Density is high.
- Reinforcement is continuous.
- Fluctuations are statistically suppressed.
- Threshold events are absorbed rather than amplified.

The world appears certain because uncertainty is integrated continuously and invisibly.

Every stable object is a choreography of micro-collapses occurring in synchrony. Nothing is ever perfectly resolved. It is continuously resolving.

Closing Transition

The classical world is not a frozen stage. It is a stabilized negotiation.

And this realization prepares us for the next structural shift.

If aggregation produces stability, what happens when aggregation compresses beyond tolerance? What happens when strain density becomes extreme? When ripple lock overloads instead of reinforcing?

The answer lies at the edges of physics. Toward alignment overload. Toward black holes.

From Persistence to Extremes

We began this chapter with a simple observation: *tables do not tremble.*

Now we understand why.

- Uncertainty did not vanish at macroscopic scale. It diluted.
- Micro-collapses did not stop. They aggregated.
- Minimal alignment packets did not disappear. They became negligible relative to total action.
- Determinism did not conquer uncertainty. It emerged from it.

Strain diluted across density. Ripple locks reinforced configuration. Aggregated stability produced predictability. Solidity became dynamic tension held collectively.

Local Determinism

But one correction must remain vivid: determinism is local.

It holds within bounded strain regimes. It depends on density, reinforcement, and constraint geometry. It is stable under ordinary conditions.

But it is not absolute.

Under extreme compression, extreme energy density, or radical constraint shifts, ripple lock can fail. Aggregation can overload. Alignment spans can compress. Threshold events can dominate instead of averaging out.

At such edges, uncertainty becomes visible again—not because it reappears, but because it was never gone.

Continuity Across Scales

The classical world is not a different universe from the quantum one. It is the same structure operating under different density conditions.

- When density supports reinforcement, certainty appears.
- When density compresses beyond tolerance, new behaviors emerge.

Transition to Extremes

The next chapter explores precisely that.

- What happens when strain does not dilute—but concentrates?
- What happens when alignment does not overlap comfortably—but overloads?
- What does collapse look like at cosmic scale?

We now move from ordinary persistence to extreme compression. From stable ripple lock to alignment overload. From the solidity of tables to the edges of spacetime itself.

Chapter 12 — Black Holes and Extreme Strain

When Stability Can No Longer Hold

So far, scale has worked in favor of stability. Strain diluted. Ripple locks reinforced. Aggregation produced persistence.

The classical world appeared certain because density supported alignment.

But what happens when density no longer stabilizes? What happens when strain does not spread—but concentrates?

In stars, gravity pulls matter inward relentlessly. In massive stars, fusion counterbalances this pull for millions or billions of years. But when fuel is exhausted, gravity wins.

Matter compresses.

Atoms collapse into neutron-dense states. Neutrons themselves crowd beyond ordinary spacing. Curvature intensifies.

Eventually, something extraordinary occurs: a black hole forms.

Black Holes as Structural Limits

For decades, black holes have been treated as exotic anomalies—regions where laws break, infinities appear, and physics fails.

But from the structural perspective we have built, we must ask:

- Is a black hole truly a violation of law?
- Or is it the revelation of structural limits?

Earlier, we saw how classical stability depends on strain dilution across density. Aggregation suppresses minimal alignment packet. Determinism emerges locally because ripple lock absorbs fluctuations.

But gravity is different. Gravity does not merely bind—it compresses.

As density increases, alignment spans shrink. Strain cannot dilute outward because curvature redirects pathways inward. Reinforcement becomes compression.

There comes a point when ripple lock can no longer stabilize through distribution. Instead of dispersing strain across networks, geometry funnels it inward.

This is alignment overload.

Under extreme gravitational constraint:

- Strain density exceeds reinforcement capacity.
- Alignment spans compress toward minimal bandwidth.
- Threshold conditions intensify rather than average out.
- Aggregation ceases to suppress uncertainty.

The very mechanism that once produced classical stability now works in reverse. Density becomes destabilizing.

What we call a black hole may not be mystical. It may be a regime where alignment overload overwhelms dilution. Where minimal alignment packet becomes relevant again at cosmic scale. Where classical determinism collapses under extreme compression.

The question is no longer whether black holes are strange. The question is whether they expose what was always true: stability is conditional.

Certainty holds only within certain strain regimes. Beyond those regimes, aggregation cannot hide uncertainty.

This chapter reframes black holes not as metaphysical voids—but as laboratories of extreme strain.

We now begin with the mechanism itself: alignment overload.

12.1 — Alignment Overload

At ordinary scale, alignment distributes. Strain flows across networks. Density supports reinforcement. Ripple lock stabilizes configuration.

But gravity introduces a different geometry.

Gravity does not merely bind particles together—it curves spacetime itself. It compresses alignment pathways. It redirects strain inward.

Compression Beyond Reinforcement

Under normal conditions, gravitational attraction is weak compared to electromagnetic interactions. At atomic and molecular scales, electromagnetic forces dominate alignment geometry.

But in massive stars, gravitational strain accumulates relentlessly.

- As mass increases, curvature intensifies.
- As curvature intensifies, strain density rises.
- As strain density rises, alignment spans compress.

What does this mean structurally?

Alignment span (ATT) reflects the duration and spatial extent over which competing strain geometries negotiate before articulation.

In dense matter, alignment is continuously reinforced. But as gravity compresses matter beyond atomic spacing, familiar reinforcement mechanisms fail.

Electrons are forced into nuclei. Protons combine with electrons into neutrons. Neutron degeneracy pressure resists further collapse—temporarily.

But if mass exceeds certain limits, even neutron reinforcement cannot stabilize configuration. Gravity overwhelms ripple lock.

This is alignment overload.

Defining Alignment Overload

Alignment overload occurs when:

- Strain density increases faster than redistribution capacity.
- Constraint geometry compresses rather than disperses.
- Alignment spans shrink toward minimal bandwidth.
- Reinforcement mechanisms saturate.

Under such conditions, aggregation no longer suppresses minimal alignment packet. Instead, minimal packet effects re-emerge at cosmic scale.

Contrast of Scales

At macroscopic human scale:

- Total action $\gg$ h.
- Minimal packet is negligible.
- Stability dominates.

Near a black hole:

- Curvature compresses alignment spans.
- Energy densities approach extreme values.
- Local action begins to approach minimal structural limits relative to geometry.
- Determinism weakens.

This does not mean Planck's constant suddenly dominates in the naive sense. It means that bandwidth constraints become structurally relevant again because alignment pathways are severely compressed.

Over-Constrained Geometry

In ordinary matter, strain redistributes outward. In black hole formation, redistribution pathways narrow. Curvature funnels strain inward. There are fewer viable articulation geometries available.

Competition collapses into singular constraint.

Alignment overload is therefore not chaos. It is over-constrained geometry.

Too much strain, too little redistribution capacity.

The very mechanism that once produced stability—density—now accelerates collapse.

Structural Continuity

This reframing clarifies something essential.

Black holes are not breakdowns of physics. They are extreme regimes of the same structural principles:

- Strain under constraint.
- Alignment under bandwidth limit.
- Threshold exhaustion under compression.

We now turn to how this overload produces large-scale threshold collapse.

Because collapse at quantum scale was articulation exhaustion. At cosmic scale, the same logic must apply.

We now examine threshold collapse under extreme curvature.

12.2 — Threshold Collapse at Cosmic Scale

We have already defined collapse in its most general form: collapse is articulation exhaustion under constraint.

At quantum scale, competing strain geometries coexist until threshold is crossed. One configuration stabilizes. Alternatives lose viability.

Now enlarge the lens.

In a collapsing massive star, gravity does not merely increase strain—it narrows viable geometries.

As density rises:

- Spatial degrees of freedom reduce.
- Energy density intensifies.
- Curvature tightens constraint geometry.
- Redistribution pathways shrink.

Competition does not disappear. It becomes constrained.

In ordinary matter, countless articulation pathways are available. Bonds can stretch, redistribute, oscillate. Energy can diffuse.

But under extreme gravitational compression, viable configurations collapse into fewer and fewer pathways.

Eventually, a critical threshold is reached. At this point, no known reinforcement mechanism—electromagnetic repulsion, neutron degeneracy pressure—can counterbalance curvature.

Threshold exhaustion occurs.

This is cosmic-scale collapse.

It is not an explosion of randomness. It is the exhaustion of viable articulation geometries under extreme constraint.

The star does not "decide" to become a black hole. Geometry loses alternatives.

Just as in quantum collapse:

- Competing configurations negotiate.
- Constraint intensifies.
- Viability narrows.
- Threshold is crossed.
- One structural regime dominates.

In black hole formation, the dominant regime is curvature-driven compression.

The event horizon emerges not as a magical boundary—but as a structural threshold. Beyond that boundary, redistribution pathways outward become non-viable. Alignment geometry is trapped within inward curvature.

From outside, it appears that matter freezes at the horizon. From inside, alignment spans compress relentlessly.

The threshold event at cosmic scale mirrors threshold events at quantum scale.

In both cases:

- Collapse does not mean annihilation.
- Collapse means narrowing of viable geometry.
- Collapse signals transition between structural regimes.

At quantum scale, collapse selects a definite outcome. At cosmic scale, collapse selects a gravitational regime where curvature dominates all other forces.

In both cases, collapse is structural inevitability under constraint intensification.

This also clarifies why black hole formation is predictable under certain mass limits. Once strain density exceeds reinforcement capacity, collapse is unavoidable. Threshold conditions determine transition.

There is nothing mystical about the inevitability of black holes beyond certain mass limits. It is geometry exhausting alternatives.

But something even more profound happens inside this regime.

As curvature intensifies, alignment spans distort. Time dilation appears. External observers perceive freezing. Internal geometry compresses. Resolution itself behaves differently.

We now turn to this phenomenon: resolution compression under extreme curvature.

12.3 — Resolution Compression

At quantum scale, we saw that articulation requires minimal alignment span (ATT). Time was reframed not as an absolute flow, but as perceived alignment duration.

Now consider what happens when curvature becomes extreme.

Near a black hole, spacetime is not merely bent—it is compressed in a specific directional geometry. Outward redistribution pathways narrow. Inward curvature dominates.

Under such conditions, alignment span distorts.

Time Dilation as Alignment Distortion

From an external observer's frame, something extraordinary appears: objects approaching the event horizon seem to slow down. Their clocks appear to tick more slowly. Eventually, they appear to freeze at the boundary.

Time dilation is not mystical. It is alignment span distortion.

If alignment span defines the duration required for relational negotiation under constraint, then curvature modifies that span.

External observers measure alignment relative to distant, less-curved geometry. Internal alignment near the horizon experiences compressed redistribution pathways.

Resolution compresses.

Structural Meaning

In ordinary regions of spacetime, strain redistributes across multiple directions. Alignment spans extend across relatively open geometry. Articulation proceeds at familiar rates.

Near extreme curvature:

- Redistribution pathways narrow.
- Alignment directions compress inward.
- External coupling weakens.
- Outward articulation visibility diminishes.

From outside, articulation appears frozen because outward alignment propagation approaches zero. From inside, relational negotiation continues—but under severe compression.

Resolution does not stop. Its visibility changes.

The event horizon therefore marks a boundary of articulation visibility, not a boundary of existence. Beyond it, alignment geometry becomes self-contained relative to external observers.

Toward Structural Limits

Time dilation near black holes reflects geometric compression of alignment span relative to external reference frames.

It is not that time itself "slows" in a mystical sense. It is that relational negotiation becomes increasingly confined.

As curvature increases further toward the classical singularity prediction, alignment span approaches structural limits. Redistribution capacity collapses. Threshold events intensify. Minimal bandwidth conditions become relevant again.

At extreme curvature, the separation between micro and macro begins to dissolve. The same structural limits that govern quantum articulation reappear at cosmic scale.

Resolution compresses toward minimal bandwidth.

Demystifying Singularity

This compression reveals something profound: the singularity predicted by classical relativity may not represent infinite density.

It may represent breakdown of classical aggregation under compressed alignment geometry.

To understand this clearly, we must demystify the singularity itself.

We now turn to what singularity really signals.

12.4 — What Singularity Really Signals

The word *singularity* has carried an almost mythical weight: infinite density, infinite curvature, breakdown of physics—a place where equations diverge and meaning collapses.

But infinity in physics is rarely a physical statement. It is almost always a structural warning.

When classical equations produce infinity, they are not revealing literal infinite quantities. They are signaling that the model has been pushed beyond its valid regime.

From the structural perspective we have built, the singularity signals something very specific: the breakdown of classical aggregation under extreme alignment compression.

Collapse of Classical Assumptions

Recall how classical stability emerged:

- Strain diluted across density.
- Ripple locks reinforced.
- Minimal alignment packet became negligible relative to total action.
- Determinism appeared smooth.

All of this depends on sufficient redistribution capacity.

In a black hole's interior, redistribution capacity collapses. Curvature funnels alignment inward. Outward articulation becomes non-viable. Density exceeds reinforcement capacity. Alignment spans compress toward minimal bandwidth.

When classical general relativity predicts infinite curvature at r=0, what it is truly indicating is this: the framework assumes continuous aggregation remains valid at all scales. But under extreme compression, aggregation fails.

Classical determinism dissolves. Minimal alignment packet can no longer be ignored.

Structural Meaning of Singularity

Singularity marks the boundary where:

- Classical continuity assumptions collapse.
- Aggregated stability fails.
- Local determinism loses coherence.
- Structural bandwidth limits reassert themselves.

In other words: singularity is not a physical point of infinite density. It is the limit of classical description under compressed strain.

This interpretation aligns with long-standing efforts in theoretical physics. Quantum gravity attempts to resolve singularities not by removing them magically, but by introducing minimal length scales or discrete structures.

From our perspective, this is natural. When alignment span approaches minimal bandwidth, classical smoothness breaks down. The infinite values in equations are not physical infinities—they are signs that alignment geometry has compressed beyond the resolution capacity of the model.

Structural Continuity

Singularity therefore signals: the collapse of classical determinism under extreme strain density.

It is not a tear in reality. It is the reappearance of minimal bandwidth limits at cosmic scale.

At quantum scale, minimal alignment packet governs articulation. At classical scale, aggregation hides it. At extreme gravitational compression, minimal bandwidth becomes relevant again.

The universe does not break. It returns to structural fundamentals.

Black holes are not mystical voids. They are laboratories where the illusion of classical certainty collapses.

And once we see this, a final insight emerges: even in such extreme regimes, resolution is never absolute. Strain cannot be fully annihilated. Residual geometry must persist. Nothing is ever fully resolved.

We now turn to the synthesis of this idea: black holes as laboratories of persistent uncertainty.

12.5 — Black Holes as Laboratories of Uncertainty

If the classical world hides uncertainty through aggregation, black holes reveal what aggregation was hiding.

They are not anomalies. They are exposure points.

In ordinary conditions, strain dilutes across density. Ripple lock stabilizes configuration. Determinism appears smooth and unbroken.

Inside a black hole regime, that smoothness fractures—not because reality becomes chaotic, but because reinforcement collapses under compression. What was averaged out becomes structurally relevant again.

Black Holes as Compression Chambers

Black holes therefore function as natural laboratories where:

- Alignment spans compress toward minimal bandwidth.
- Redistribution pathways collapse inward.
- Threshold behavior dominates over dilution.
- Classical determinism reaches its limit.

The illusion of certainty dissolves.

Consider what happens at the event horizon. From the outside, information appears frozen. From inside, geometry continues evolving under extreme constraint. The boundary marks a shift in articulation visibility—not annihilation of structure.

Even Hawking radiation suggests something profound: black holes are not static tombs. They emit. They evaporate. They transform.

If strain were fully resolved inside, nothing would leak. But persistence remains.

Persistence of Residue

This supports a deeper principle: nothing is ever fully resolved.

Even under extreme collapse, residual strain geometry persists. Alignment cannot eliminate structure entirely—it can only compress and transform it.

Black holes are therefore not endpoints. They are compression chambers.

They test the limits of:

- Alignment bandwidth.
- Aggregated determinism.
- Classical geometry.

They demonstrate that:

- Certainty was conditional.
- Aggregation was local.
- Stability was density-dependent.

Under sufficient strain, even the most rigid-seeming regimes reveal negotiation again.

Structural Continuity Across Scales

This aligns seamlessly with everything established so far:

- At minimal scale, uncertainty is visible.
- At ordinary scale, uncertainty is diluted.
- At extreme scale, uncertainty resurfaces under compression.

The universe is structurally consistent across regimes.

Black holes are not breaks in law. They are structural mirrors.

They show us that certainty was never absolute—it was scale-dependent ripple lock.

Toward Persistence

And once we understand this, the next insight becomes unavoidable:

If even extreme collapse cannot fully resolve strain, if singularity signals boundary rather than termination, if residual geometry persists—then nothing in reality is ever fully finished.

Resolution is always partial. Uncertainty is layered. Persistence carries residue.

The universe does not reach final stillness. It transitions between regimes of negotiation.

We now move toward that realization: From extreme compression to incomplete resolution. From cosmic overload to the truth that nothing is ever fully resolved.

Chapter 13 — Nothing Is Ever Fully Resolved

The Myth of Finality

Human intuition loves completion. A problem solved. A decision made. A structure built. A star collapsed.

We speak of endings as if they are absolute: The wavefunction collapses. The system stabilizes. The star becomes a black hole. The matter reaches singularity.

We imagine final states.

But have we encountered a single example of true finality?

- When a wavefunction collapses, articulation occurs. Yet the resulting configuration immediately participates in new relational geometry.
- When a solid stabilizes, micro-collapses continue beneath its surface.
- When a star collapses into a black hole, curvature evolves, Hawking radiation emerges, geometry transforms.

At every scale, what we call "resolution" is merely local exhaustion within a boundary—not elimination of strain.

Resolution does not mean strain disappears. It means competing geometries lose viability under a given constraint set. But change the boundary—and new competition emerges.

Structural Pattern

This pattern has been consistent across every chapter:

- At quantum scale, collapse is threshold exhaustion—not annihilation.
- At classical scale, stability is aggregated negotiation—not stillness.
- At extreme cosmic scale, singularity signals model breakdown—not absolute infinity.

So what does this tell us?

It tells us that full resolution—absolute exhaustion of all strain across all scales—is structurally impossible.

Why? Because alignment bandwidth is finite. Constraint geometry is layered. Redistribution is never total.

Every articulation produces new boundaries. Every boundary produces new deltas. Resolution generates residue. And that residue propagates.

The universe does not terminate its negotiations. It shifts them.

We mistake stabilization for completion because our perception operates within bounded regimes. When a system becomes stable relative to our scale, we label it "resolved."

But relative stability is not universal closure. Nothing ever collapses into absolute stillness.

If it did, motion would cease. Time would vanish. Evolution would stop.

Yet dynamism persists. Stars burn. Galaxies collide. Particles fluctuate. Thoughts arise.

The myth of finality collapses under structural inspection. Resolution is never total. It is bounded alignment within layered uncertainty.

This chapter completes Part IV by making that principle explicit: nothing is ever fully resolved.

Not quantum systems. Not classical objects. Not black holes.

Strain persists. And it is precisely this persistence that makes reality dynamic.

13.1 — Residual Strain

Every collapse leaves something behind.

At quantum scale, collapse selects an articulation. One configuration stabilizes under threshold exhaustion. But does that eliminate all competing geometry? No.

It eliminates competing geometry within that boundary condition. The articulated state immediately enters new relational contexts. It interacts. It propagates. It becomes part of larger constraint geometry. Strain redistributes again.

Residual delta remains.

Collapse as Reconfiguration

When a wavefunction collapses under measurement, one outcome is selected. But the global relational geometry—the environment, the measuring apparatus, the broader system—continues to carry redistributed strain. Alignment has shifted, not ended.

Collapse is reconfiguration, not annihilation.

Similarly, in classical systems, when equilibrium is reached, we speak of stability. But equilibrium is not absence of strain. It is balanced strain under current constraints.

Atoms vibrate. Bonds stretch microscopically. Thermal fluctuations persist.

Residual strain is continuously present.

In fact, if residual strain were absent, motion would cease. There would be no thermal energy, no dynamism, no entropy flow.

Residual strain is not a defect. It is structural necessity.

Cosmic Scale

Extend this to cosmic scale.

When a star collapses into a black hole, gravitational collapse exhausts many articulation pathways. But does strain vanish? No.

Curvature intensifies. Hawking radiation emerges. Quantum effects near the horizon persist.

The collapse generates new boundary conditions—and therefore new relational negotiation.

Even the classical singularity is not a final state. It signals breakdown of description under compressed strain—not elimination of geometry.

Residual strain persists across every scale.

Structural Statement

This leads to a powerful structural statement: Resolution always creates new boundaries. New boundaries generate new deltas.

There is no global exhaustion of strain. There is only local articulation under bounded constraint.

Analogy

Consider a simple analogy.

You solve a problem. That solution creates new circumstances. New constraints arise. New questions emerge.

Resolution generates context. Context generates strain. Strain generates negotiation.

The universe behaves similarly.

Every articulation narrows some pathways and opens others. Residual strain propagates across scale.

This propagation explains why dynamism persists.

If full exhaustion were possible—if strain could be globally eliminated—the universe would freeze.

But because alignment bandwidth is finite and geometry layered, redistribution is never complete.

Residual delta ensures motion continues.

The persistence of strain is therefore not accidental. It is structural.

13.2 — Layered Uncertainty Across Scales

Residual strain does not exist at a single level. It exists across layers.

This is one of the most important realizations of the entire book: uncertainty is layered.

It is not confined to quantum mechanics. It is not eliminated at classical scale. It is not destroyed at cosmic scale.

It is stratified.

Multi-Scale Visibility

At quantum scale, uncertainty is visible because alignment bandwidth is minimal. Competing geometries coexist. Collapse selects articulation locally, but residual strain persists in relational context.

Uncertainty here is explicit.

At classical scale, micro-uncertainty becomes hidden by aggregation. Trillions of articulation events overlap and average. Stability emerges. Determinism appears smooth.

But the underlying micro-negotiation continues. Uncertainty here is diluted, not removed.

At cosmic scale, extreme compression re-exposes structural limits. Classical aggregation fails. Alignment spans compress. Singularity signals breakdown of continuity assumptions.

Uncertainty here reappears as boundary instability.

Unified Layers

Now step back.

These are not separate domains with separate rules. They are layers of the same structural system.

Each layer operates under:

- Its own scale of density.
- Its own bandwidth relevance.
- Its own redistribution capacity.
- Its own constraint geometry.

Uncertainty shifts visibility depending on scale. But it never disappears.

Analogy

Think of it like geological layers in the Earth.

At the surface, soil seems stable. Beneath it lie tectonic plates. Beneath them, molten dynamics. The surface appears calm because deeper forces are distributed and balanced.

But the deeper layers are always active.

Reality functions similarly.

At any moment:

- Quantum layers negotiate micro-alignment.
- Classical layers aggregate reinforcement.
- Cosmic layers evolve under curvature.

- Biological systems process strain internally.
- Psychological systems negotiate cognitive deltas.

Each layer contains its own uncertainty regime. None eliminates the others.

Structural Consequence

Layered uncertainty explains why no final equilibrium is ever reached.

When one layer stabilizes, deeper layers continue negotiation. When deeper layers compress, upper layers feel instability.

This is why stability is always conditional. It is not global—it is layered and local.

Even in human experience, this pattern holds. You may resolve a conscious conflict. Yet subconscious tensions persist. Social structures stabilize, yet underlying economic or psychological strain accumulates.

Resolution at one layer does not eliminate uncertainty at another.

Layered uncertainty is structural inevitability in any system with:

- Finite bandwidth,
- Multiple interacting scales,
- Non-total redistribution capacity.

This realization dissolves the myth of universal closure.

There is no ultimate stillness because layers remain active.

Uncertainty is not a singular phenomenon—it is stratified.

And this layering ensures that persistence is never passive. Stability survives through partial alignment.

13.3 — Persistence Through Partial Alignment

If nothing is ever fully resolved, why does anything persist at all? Why do atoms endure? Why do ecosystems stabilize? Why do galaxies maintain structure for billions of years?

The answer is simple—but subtle: persistence does not require total resolution. It requires sufficient alignment.

Partial alignment is enough.

Structural Meaning

Alignment occurs when competing strain geometries negotiate within bandwidth limits and produce a stable configuration under current constraints.

But "stable" does not mean strain-free. It means strain is balanced enough to avoid runaway collapse.

Atoms: Electrons do not sit motionless around a nucleus. They exist in relational probability distributions—structured strain geometry. Yet atoms persist because alignment between electromagnetic attraction and quantum bandwidth constraints produces a bounded configuration. The atom is not fully resolved. It is sufficiently aligned.

Planets: Gravitational pull draws them inward. Inertial motion carries them forward. The orbit is not absence of force—it is dynamic balance between competing geometries. Remove one element and collapse occurs. Persistence arises from continuous negotiation.

Macroscopic objects: They maintain shape not because strain disappears, but because micro-articulations reinforce one another strongly enough to prevent runaway divergence. Thermal fluctuations exist. Micro-strain persists. But alignment strength exceeds destabilizing impulses.

Universal Principle

This principle extends universally:

- Biological organisms persist through metabolic balancing of internal and external strain.
- Ecosystems persist through dynamic equilibrium of competition and cooperation.
- Social systems persist through negotiated stability between conflicting interests.

None of these systems eliminate strain. They regulate it.

Partial alignment is active maintenance. It requires continuous micro-resolution.

If alignment ever became total—if no competing geometry remained—motion would cease. Persistence would freeze into stillness.

But finite bandwidth prevents total exhaustion. Residual strain ensures ongoing adjustment.

This is why equilibrium is never static. It is dynamic equilibrium. A stabilized negotiation. A ripple lock maintained across scale.

Persistence is therefore not a state. It is a process. It is alignment sustained through ongoing articulation.

And because alignment is always partial, reality remains dynamic.

13.4 — Why Reality Remains Dynamic

If full resolution were possible, reality would eventually come to rest. All strain would exhaust. All competing geometries would vanish. All motion would cease.

The universe would freeze into perfect equilibrium.

But this has never occurred— Not at quantum scale. Not at classical scale. Not at cosmic scale.

Structural Drivers of Dynamism

Why does reality remain dynamic? Because finite bandwidth guarantees incomplete exhaustion.

1. **Finite alignment span.** Resolution cannot occur instantaneously across infinite relational networks. Alignment is local and bounded. Each articulation resolves only within its immediate constraint geometry.
2. **Layered interaction.** Systems exist within systems. A stable molecule exists within a cell. A stable cell exists within an organism. An organism exists within an ecosystem. Each layer introduces new relational geometry. Stability at one layer does not imply stability at another.
3. **Residual strain propagation.** Every articulation redistributes strain into adjacent layers. This redistribution creates new deltas elsewhere. Strain does not vanish—it migrates.
4. **Entropy and redistribution**. Entropy is often misunderstood as disorder. Structurally, it reflects redistribution of strain across accessible configurations. As systems evolve, they explore relational possibilities within bandwidth limits. Entropy growth ensures continued reconfiguration.

Together, these mechanisms guarantee that reality cannot achieve absolute stillness.

Examples

- **Stars:** Fusion stabilizes gravitational strain for billions of years. But fuel depletion changes boundary conditions. Collapse follows. New structures emerge.
- **Matter:** Atoms are stable under ordinary conditions. But high energy input alters alignment. Ionization occurs. Chemical reactions proceed. Bonds break and reform.
- **Cosmic scale:** Even black holes evaporate through Hawking radiation. Compression does not create eternal stillness. Geometry evolves.

Time as Negotiation

The universe persists because it never reaches complete exhaustion of strain.

Dynamism is not a side effect. It is structural inevitability under finite bandwidth and layered constraint geometry.

This reframes time itself. Time is not a flowing substance. It is the perceptual measure of ongoing alignment negotiation.

If negotiation ceased, time would lose meaning. But negotiation cannot cease because:

- Alignment is partial.
- Redistribution is incomplete.
- Residual strain propagates.

Reality remains dynamic because full resolution is structurally impossible.

Final Synthesis

And this brings us to the final synthesis of this chapter: Incompleteness is not failure. It is the engine of continuity.

We now formalize this as a structural law.

13.5 — The Structural Law of Incompleteness

We are now ready to state it clearly. Not as poetry. Not as philosophy. But as structure.

No system ever achieves total resolution. Not quantum systems. Not classical systems. Not cosmic systems.

Resolution is always local, bounded, and conditional.

This is not a limitation of our models. It is a property of reality.

Why Resolution Is Always Partial

- Alignment is finite.
- Bandwidth is limited.
- Constraint geometry is layered.
- Redistribution is never global.

Every articulation exhausts strain within a boundary—but that boundary itself exists within a larger relational field.

Thus, resolution at one scale generates new strain at another.

This is the structural law of incompleteness.

Expressed Simply

Total exhaustion of strain across all layers is impossible under finite alignment bandwidth.

- What appears as completion is stabilization within a

regime.
- What appears as finality is boundary-dependent exhaustion.
- What appears as certainty is layered partial alignment.

Unified Across Scales

This law unifies everything we have explored:

- Quantum collapse is partial articulation.
- Classical determinism is aggregated suppression.
- Black hole singularity signals breakdown of aggregation—not universal termination.
- Persistence across scales depends on ongoing negotiation.

Nothing ever reaches absolute stillness. The universe does not finish resolving itself. And that is precisely why it continues.

If full resolution were possible, evolution would stop. No stars would burn. No atoms would vibrate. No thoughts would arise.

Continuity exists because incompleteness persists.

Reframing Uncertainty

This reframes uncertainty itself.

Uncertainty is not ignorance. It is structural residue. It is the trace of incomplete alignment. It is the space where negotiation continues.

Incompleteness is not a flaw in reality. It is the engine of continuity.

It allows:

- Motion.
- Transformation.

- Adaptation.
- Emergence.

The universe survives because it never fully closes.

Every collapse leaves residue. Every stabilization leaves layers. Every certainty contains suppressed strain.

And because suppression is never elimination, dynamism remains.

This part closes with clarity:

- Certainty was conditional.
- Collapse was bounded.
- Singularity was structural limit.
- Resolution was always partial.

Nothing is ever fully resolved.

Which means uncertainty is not an exception in reality. It is the foundation of its continuity.

From Cosmic Incompleteness to Human Uncertainty

We have now followed uncertainty across scales: From quantum articulation to classical aggregation, from gravitational compression to structural incompleteness.

At every stage, one conclusion remained consistent: nothing is ever fully resolved.

- Collapse is local.
- Stability is conditional.
- Determinism is scale-bound.
- Singularity is a boundary of description.

Residual strain persists. Layered uncertainty survives. Alignment is always partial.

This is not a poetic metaphor. It is structural necessity under finite bandwidth and layered constraint geometry.

The universe continues because it never reaches final exhaustion.

Human Dimension

Now pause.

If incompleteness governs matter, if persistence depends on partial alignment, if residual strain drives dynamism—why would the human mind be exempt?

Why would cognition achieve total resolution when atoms cannot? When stars cannot? When spacetime itself cannot?

Doubt, hesitation, indecision, curiosity—these are not psychological weaknesses. They are expressions of structural incompleteness within conscious systems.

Certainty in thought functions like classical determinism: it is aggregated suppression of underlying micro-negotiation.

When strain compresses—emotionally, intellectually, socially—uncertainty reappears visibly.

Just as black holes reveal structural limits of classical physics, crises reveal structural limits of psychological stability.

The same geometry persists. Only the layer changes.

We now move into the human field: From cosmic strain to cognitive strain. From singularities of curvature to singularities of belief. From structural incompleteness in physics to lived incompleteness in experience.

The framework does not change. The scale does.

And in that shift, uncertainty will reveal its most intimate dimension.

Interlude — Nothing is ever fully resolved

We began this part with a simple observation: *the classical world feels certain.*

We now know that certainty was never absolute. It was structural dominance under favorable conditions.

Core Realizations

- Classical certainty emerges from strain dilution across density. Micro-uncertainty does not vanish—it becomes statistically suppressed.
- Large-scale ripple lock stabilizes systems through collective reinforcement of countless micro-articulations.
- Determinism is not universal. It is local to bounded strain regimes.
- Solidity is dynamic tension, not frozen stillness. Stability is maintained through continuous micro-resolution.
- Extreme gravitational compression produces alignment overload. Density stops stabilizing and begins constraining.
- Black hole formation is threshold exhaustion at cosmic scale, not metaphysical rupture.
- The event horizon marks a boundary of articulation visibility, not annihilation of structure.
- Singularity signals breakdown of classical aggregation, not infinite physical reality.
- Minimal bandwidth limits re-emerge under extreme

compression.

- Collapse does not eliminate strain—it reorganizes it.
- Residual strain persists after every articulation.
- Uncertainty is layered across scales—quantum, classical, cosmic.
- Stability survives through partial alignment, not total resolution.
- Full exhaustion of strain is structurally impossible under finite bandwidth.
- Reality remains dynamic because incompleteness is fundamental.
- The universe continues because it never fully resolves itself.

This part closes with structural maturity: Resolution is always local. Certainty is always conditional. Finality is always boundary-dependent.

Nothing is ever fully resolved.

And this realization changes everything.

If incompleteness governs matter and spacetime, then uncertainty must also shape cognition.

The next part shifts the lens:

- From physics to perception.
- From cosmic incompleteness to human experience of doubt.

The structure remains the same. Only the scale changes.

Part V — Uncertainty in the Human Field

"The mind is not disturbed by uncertainty. It is structured by it."

Overview

Until now, uncertainty has been examined across physical scales.

- At the quantum level, it appeared as unresolved geometry.
- At the classical level, it was diluted through aggregation.
- At cosmic extremes, it resurfaced under compression.

But one question now becomes unavoidable: If incompleteness is structural in reality, what does it mean for the human mind?

This part shifts the lens from external geometry to internal experience.

Here, uncertainty is no longer abstract. It becomes doubt, hesitation, curiosity, fear, creativity, belief, identity.

The human mind is not a spectator of uncertainty. It is a participant within it.

Cognition as Structural Negotiation

- Cognition operates under finite bandwidth.
- Perception filters layered strain.
- Belief systems act as ripple locks.
- Crises resemble alignment overload.

Uncertainty in physics was never mystical. Uncertainty in human life is not weakness.

Both arise from the same principle: Alignment is always partial. Resolution is always local.

And consciousness lives inside that negotiation.

Chapter 14 — Psychological Uncertainty

Why the Mind Cannot Rest

We have seen that nothing in the physical universe reaches total resolution. Collapse is local. Stability is conditional. Incompleteness is structural.

Now turn inward.

Why does the mind behave the same way?

Why does doubt arise even after a firm decision? Why does certainty weaken over time? Why do beliefs that once felt unshakable suddenly tremble under new information?

Why can the mind never fully rest?

We often interpret psychological uncertainty as weakness—indecision, overthinking, lack of confidence.

But what if uncertainty in the mind is not a flaw? What if it is structural—just like uncertainty in physics?

The brain processes competing possibilities constantly. Every thought is a negotiation between memory, perception, expectation, emotion, and social influence. Attention—like alignment bandwidth—is finite. It cannot evaluate every possibility simultaneously.

So the mind collapses locally. It chooses. It decides. It forms belief.

But just as in quantum articulation, that collapse is bounded. Residual cognitive strain remains in memory, in subconscious layers, in alternative interpretations.

No decision exhausts all possible thought pathways. No belief eliminates every competing narrative.

The mind stabilizes through partial alignment. And because alignment is partial, doubt returns.

Convictions and Context

Consider how even strong convictions evolve. You once believed something absolutely. Years later, you reinterpret it. Was your earlier certainty false? Or was it sufficient alignment under earlier constraints?

Cognitive stability, like classical determinism, depends on context and reinforcement. When context shifts, suppressed strain resurfaces.

This is why crises feel destabilizing. Why contradictory evidence causes discomfort. Why identity conflicts produce anxiety.

The mind cannot rest because it is not designed for total resolution. It is a dynamic alignment system operating under finite bandwidth.

Just as physical systems never achieve perfect stillness, the mind never achieves perfect closure.

Psychological uncertainty is not accidental. It is the cognitive expression of structural incompleteness.

This chapter examines that structure. We begin with the most fundamental layer: cognitive strain.

14.1 — Cognitive Strain

At the physical level, we called it *delta.* At the psychological level, we call it *tension.*

But structurally, it is the same phenomenon: competing geometries under finite bandwidth.

The mind is not a single voice. It is a negotiation field.

Memory pulls in one direction. Emotion pulls in another. Social expectation adds pressure. Future prediction introduces uncertainty.

Each of these is a cognitive strain vector.

Decision as Collapse

When you face a decision, you are not choosing between two abstract options. You are navigating competing internal geometries:

- One pathway preserves comfort.
- Another promises growth.
- One protects identity.
- Another challenges it.

These pathways cannot all be articulated simultaneously with equal intensity because attention is finite.

Attention is psychological alignment bandwidth.

You cannot fully evaluate every possibility at once. The mind must compress. It must prioritize. It must select.

This selection feels like decision. But structurally, it is local threshold collapse.

You reach a point where cognitive strain exceeds bandwidth tolerance. The mind stabilizes by articulating one pathway as *the decision.*

Relief follows—not because all strain is gone, but because one configuration has temporarily exhausted competing narratives within that boundary.

Residual strain remains.

That is why second thoughts occur. That is why doubt creeps in later. That is why even after making the "right" choice, unease can persist.

The decision resolved immediate overload. It did not eliminate all competing geometry.

Example

Consider a simple example: you accept a job offer.

For a moment, clarity emerges. The tension reduces. But later, alternative possibilities reappear in thought. *What if I had chosen differently? Was this truly optimal?*

The mind is revisiting residual pathways.

Cognitive strain never vanishes completely because the decision only resolved strain relative to immediate constraints. New information, new emotion, new context reactivates suppressed geometry.

This continuous negotiation explains why certainty fades over time.

Just as classical determinism is local to strain regime, psychological certainty is local to context regime. When context changes, alignment recalculates.

Structural Inevitability

Cognitive strain is therefore not pathology. It is structural inevitability in any system that:

- Has finite processing bandwidth.
- Operates within layered constraint environments.
- Stores memory of alternative geometries.

The mind cannot permanently silence alternatives. It can only stabilize temporarily.

And because stabilization requires reinforcement, humans build narratives.

Narratives act like classical aggregation. They suppress visible uncertainty.

We now examine how narrative amplification strengthens psychological ripple lock—and why suppression is never elimination.

14.2 — Narrative Amplification

If cognitive strain is the raw geometry of uncertainty, narrative is its stabilizer.

The mind does not tolerate open negotiation for long. Continuous ambiguity is bandwidth-expensive. So it builds stories.

Stories compress complexity. They select one articulation and reinforce it repeatedly until it feels natural, inevitable, even "true."

This is narrative amplification.

When you make a decision, the mind does not merely store the outcome. It constructs a justification:

- "This was the logical choice."
- "This aligns with who I am."
- "This is what successful people do."
- "This is what I always believed."

Each repetition strengthens alignment. Competing geometries weaken—not because they disappear, but because attention reinforces one pathway more than others.

Narrative acts like classical aggregation. Just as trillions of micro-collapses average into determinism, repeated cognitive reinforcement averages into belief stability.

The more a narrative is repeated—internally or socially—the more it becomes a psychological ripple lock.

Collective Reinforcement

This is why social validation is powerful. When others confirm your belief, their narratives synchronize with yours. Alignment spans overlap. Reinforcement multiplies. Collective storytelling creates large-scale cognitive aggregation.

But amplification is not elimination. Suppressed strain remains beneath the surface.

Consider ideological certainty. A person may defend a belief fiercely, recite its justifications flawlessly, reject opposing views aggressively. Yet under sufficient strain—contradictory evidence, emotional shock, social shift—cracks appear.

Why? Because the narrative reinforced one articulation, but residual geometry was never destroyed. It was suppressed.

Limits of Narrative

Narrative amplification reduces visible uncertainty by:

- Repetition,
- Emotional reinforcement,
- Social synchronization,
- Identity integration.

But it cannot exhaust all competing possibilities.

This explains why doubt resurfaces during crisis. When bandwidth is overloaded—through stress, loss, or contradiction—the suppressed geometry regains viability. The ripple lock weakens. The story trembles.

Narrative is therefore not deception. It is a stabilizing mechanism. It helps the mind function under finite bandwidth.

But it has limits. Over-amplified narratives become brittle. Like classical determinism pushed beyond its strain regime, rigid belief systems collapse when compression exceeds reinforcement capacity.

The stronger the suppression, the sharper the eventual fracture.

This leads us to the deepest anchor in psychological geometry: identity.

14.3 — Identity as Geometric Anchor

If narrative stabilizes thought, identity stabilizes the narrator.

Identity is not merely a label. It is a geometric anchor in cognitive space.

"I am this kind of person." "I belong to this group." "I believe in this system." "I stand for these values."

Each of these statements creates a center of alignment. Around this center, thoughts cluster. Memories reorganize. Decisions align. Narratives reinforce.

Identity acts like large-scale ripple lock in the psychological field. Just as classical objects maintain stability through dense reinforcement, identity maintains psychological stability through clustered beliefs and repeated self-confirmation.

Anchoring and Efficiency

This clustering reduces cognitive strain. If every decision had to be evaluated from zero, bandwidth would collapse. Identity compresses complexity by pre-filtering options:

- "This is consistent with who I am."
- "That is not."

The mind saves energy through anchoring.

Vulnerability of Anchors

But anchoring has consequences.

When identity becomes tightly reinforced, competing geometries threaten more than a single belief—they threaten the anchor itself.

If a narrative cracks, discomfort arises. If identity cracks, existential anxiety emerges.

Why? Because identity is the organizing geometry of the mind.

It integrates:

- Memory,
- Social role,
- Emotional investment,
- Moral framework,
- Future projection.

Threaten identity, and alignment span compresses sharply. This resembles alignment overload at psychological scale.

Identity Under Threat

Under intense identity threat:

- Attention narrows.
- Defensive narratives amplify.
- Contradictory evidence is rejected.
- Emotional intensity increases.

The mind is attempting to preserve anchor stability.

Consider ideological conflict. When a belief is peripheral, debate feels intellectual. When a belief is identity-linked, disagreement feels personal—even existential.

The geometry has shifted from surface strain to anchor strain. The emotional charge is not about facts. It is about structural stability.

Identity as Conditional Stability

Identity, like classical determinism, functions within bounded regimes. It holds under ordinary strain. But under compression—social upheaval, betrayal, failure, radical new evidence—the anchor can destabilize.

When that happens, uncertainty feels deeper than cognitive doubt. It feels like loss of self.

This is why identity transitions are destabilizing: career shifts, religious transformation, political reorientation, cultural displacement.

These are not simple belief updates. They are geometric reconfigurations.

Identity is therefore both stabilizer and vulnerability. It allows persistence through partial alignment. But because alignment is never total, identity can never eliminate underlying uncertainty.

The anchor itself rests on layered incompleteness.

Which brings us to the emotional dimension: why doubt feels existential, and why questioning oneself can feel like structural collapse.

We now turn to that intensity.

14.4 — Why Doubt Feels Existential

Not all doubt feels the same.

Some doubts are light: *What should I eat? Which route should I take?* These create minor cognitive strain. Alignment span negotiates quickly. Residual uncertainty is negligible.

But other doubts feel destabilizing: *Am I in the right career? Is my belief system true? Who am I without this identity? What if everything I trusted is wrong?*

These do not feel like surface negotiation. They feel like structural tremors.

Anchor Geometry and Compression

Why? Because doubt becomes existential when it threatens the anchor geometry.

When uncertainty touches identity, the ripple lock compresses. Recall what we observed in black holes: under extreme compression, redistribution pathways narrow. Alignment overload occurs.

Psychological doubt behaves similarly.

When a belief loosely attached to identity is challenged, cognitive strain distributes easily. Alternative narratives can be considered without panic.

But when a core identity belief is challenged:

- Bandwidth narrows.
- Emotional charge increases.
- Defensive narratives amplify.
- Contradictory geometry is rejected aggressively.

The mind experiences anchor destabilization.

Groundlessness

Existential doubt is not just "not knowing." It is fear of losing organizing geometry.

Identity gives coherence to memory, purpose to action, direction to future projection. Remove that coherence and the mind faces open negotiation without stable reference.

That openness feels like groundlessness. Groundlessness feels like danger.

This is why doubt feels existential—not because uncertainty itself is dangerous, but because anchor instability threatens psychological persistence.

Humans evolved to prefer stable alignment. Stability conserves energy, reduces cognitive load, and enables coordinated social interaction.

When anchor stability shakes, the system perceives risk. Emotion intensifies. Anxiety, anger, denial, or withdrawal may follow.

But structurally, this is not weakness. It is the natural response of a partial alignment system encountering compression.

Existential doubt exposes structural incompleteness directly. It reveals that identity, like classical determinism, was conditional.

Rigid identity increases brittleness. Flexible identity increases adaptability.

The intensity of existential doubt therefore signals proximity to anchor geometry—not proximity to truth or falsity.

Critical Realization

Doubt is not a malfunction. It is the felt experience of structural incompleteness at identity scale.

And if incompleteness is universal, then psychological stability must depend not on eliminating doubt—but on managing partial alignment.

We now move to the final synthesis of this chapter: the mind as a partial alignment system.

14.5 — The Mind as Partial Alignment System

We now arrive at the central psychological realization of this book:

The mind does not seek total truth. It seeks sufficient alignment.

It is not a certainty machine. It is a stabilization system.

Just as physical systems persist through partial alignment, the mind survives through bounded coherence.

Conditions for Stability

A belief system feels stable when:

- Cognitive strain is manageable.
- Identity remains coherent.
- Narrative reinforcement is sufficient.
- Social validation overlaps with internal geometry.

Under these conditions, uncertainty becomes background noise.

But notice what is absent: total resolution.

No belief system eliminates all competing interpretations. No identity exhausts all alternative self-configurations. No narrative explains every anomaly.

Residual strain remains. Yet the system functions.

Why Partial Alignment Is Enough

The mind does not require absolute certainty to act. It requires coherence within bandwidth tolerance.

This is why humans can operate with incomplete information, evolving understanding, and shifting beliefs.

Psychological stability is not built on perfection. It is built on resilience within incompleteness.

When strain rises beyond tolerance—through crisis, contradiction, or trauma—the system destabilizes.

But destabilization is not failure. It is recalibration.

Just as physical collapse reorganizes geometry, psychological collapse reorganizes identity and narrative.

Growth often follows destabilization precisely because suppressed geometry becomes visible.

Flexibility and Integration

A rigid system shatters. A flexible system adapts.

The healthiest psychological systems are not those that eliminate doubt. They are those that integrate it.

They maintain anchor flexibility. They allow partial alignment to evolve. They tolerate layered uncertainty without demanding total closure.

Reframing Key Concepts

This reframes confidence: Confidence is not absence of doubt. It is functional coherence despite doubt.

This reframes intelligence: Intelligence is not perfect resolution. It is effective navigation within incomplete geometry.

This reframes maturity: Maturity is recognition that certainty is conditional.

The mind, like the universe, persists because it never fully resolves itself.

Residual strain drives curiosity. Incomplete alignment drives growth. Layered uncertainty sustains evolution of thought.

The goal is not to eliminate uncertainty. It is to manage alignment within it.

Transition to Collective Scale

And once we understand this at individual scale, a larger question emerges:

What happens when millions of partial alignment systems interact? When identities cluster into ideologies? When narratives scale into culture?

Uncertainty does not disappear at collective level. It amplifies.

We now widen the lens from the individual mind to the social field— From personal uncertainty to collective strain.

From the Individual Mind to the Collective Field

We began this chapter by asking why the mind cannot rest.

We now know the answer. It cannot rest because it is built on partial alignment.

Cognitive strain never vanishes. Narratives amplify but do not eliminate uncertainty. Identity stabilizes but cannot guarantee permanence. Doubt feels existential because anchors are conditional.

The mind survives not by finishing negotiation—but by managing it.

Structural Continuity

This realization carries a subtle but powerful shift: Uncertainty is not a psychological defect. It is structural continuity expressed in consciousness.

- Every belief you hold is a stabilized ripple lock.
- Every decision you make is a local threshold collapse.
- Every crisis you face is alignment overload at identity scale.

The geometry is the same as physics. Only the layer has changed.

Scaling the Pattern

But something larger now becomes visible.

If one mind operates under finite bandwidth and layered uncertainty, what happens when millions of such minds interact?

When identities cluster. When narratives synchronize. When belief systems scale into institutions.

Collective systems amplify the same structure:

- Aggregated certainty emerges.
- Narrative reinforcement multiplies.
- Identity anchors harden.
- Suppressed strain accumulates beneath surface stability.

Just as classical determinism hid quantum uncertainty, social stability hides collective strain. And just as extreme compression exposed structural limits in black holes, social compression exposes structural limits in cultures and institutions.

The geometry scales. The pattern persists.

We now move from the psychological field to the collective field. From individual incompleteness to social instability. From private doubt to shared uncertainty.

The structure has not changed. Only its scale has grown.

Chapter 15 — Spiritual Significance

Why the Unknown Feels Larger Than Us

Stand beneath a night sky far from the city. No narrative. No ideology. No identity performance. Just vastness.

Something happens in that moment.

The mind becomes quiet—not because questions are answered, but because they expand beyond the boundaries of immediate control. Awe arises.

The Dual Nature of Uncertainty

Why does uncertainty sometimes feel frightening, and at other times sacred? Why does the unknown before death feel heavy, yet the unknown before the cosmos feel profound? Why does silence in a temple feel different from silence in confusion?

The structure is the same. The interpretation changes.

Earlier, we saw that the mind cannot achieve total resolution. Identity stabilizes through partial alignment. Narrative suppresses visible doubt. Certainty is local.

But occasionally, the mind encounters scale beyond its anchor geometry: the night sky, the ocean, birth, death, infinity.

In such moments, bandwidth recognizes its own limit. The anchor compresses—but instead of resisting, something softens.

This softening is what humans have historically called spiritual.

Spirituality as Structural Response

Spirituality, at its structural core, may not be about supernatural explanation. It may be about encountering incompleteness without panic.

When identity meets something larger than its organizing geometry, two responses are possible:

- Resistance — which amplifies anxiety.
- Surrender — which reduces alignment overload.

Awe arises when resistance drops. Uncertainty feels sacred when it is no longer treated as a problem to solve.

Consider the difference between existential panic and contemplative stillness. In both cases, the mind confronts incompleteness. But in panic, it demands resolution. In contemplation, it allows partial alignment.

The unknown feels larger than us because it is—not in a mystical sense, but in structural scale.

The universe operates across layered geometries beyond individual bandwidth. The mind senses its finiteness relative to that scale. That recognition can destabilize identity—or deepen humility.

Humility is not weakness. It is accurate scale awareness.

Spiritual traditions across cultures speak of surrender, acceptance, presence, silence. Stripped of ritual and doctrine, these may be techniques for stabilizing the mind under structural incompleteness.

The sacred may not lie in certainty. It may lie in the recognition that certainty was never absolute.

This chapter explores spirituality not as belief—but as structural response to incompleteness.

We begin with the first layer: why uncertainty, when approached without resistance, feels sacred.

15.1 — Why Uncertainty Feels Sacred

Uncertainty does not always feel threatening. Sometimes it feels overwhelming. Sometimes it feels destabilizing. And sometimes—it feels sacred.

The structure underneath all three experiences is the same: encounter with incompleteness. The difference lies in how the mind responds to bandwidth limits.

Scale Awareness

When the mind confronts something beyond its capacity to fully articulate—vastness, death, infinity, silence—it becomes aware of its own boundary.

Attention reaches outward. Narrative fails to compress adequately. Identity cannot fully anchor the scale of what is perceived. Bandwidth recognizes its own finiteness.

That recognition produces cognitive compression.

If the mind resists this compression—trying to force explanation, demand certainty, defend identity—anxiety emerges.

But if resistance drops, something else happens. The mind relaxes into partial alignment. It stops attempting total resolution.

That relaxation produces awe.

Awe is the emotional signature of scale-awareness without defensive contraction. It is not ignorance. It is humility before layered structure.

Sacredness as Spaciousness

Consider standing at the edge of a vast ocean. You cannot cognitively process every wave, every current, every depth. The system is too large relative to your bandwidth.

Yet you do not feel attacked. You feel small—but not annihilated.

Why? Because identity temporarily softens. The anchor loosens. Narrative amplification quiets.

In that quiet, uncertainty becomes spacious rather than threatening.

This spaciousness is what many traditions call sacred. Not because it proves a supernatural claim, but because it reveals structural proportion.

You experience yourself as a partial alignment system inside a vastly layered universe.

The ego contracts less. The need for total resolution diminishes. The mind stabilizes without closure.

Sacredness as Alignment

Sacredness, in this sense, is cognitive alignment with incompleteness.

It is the recognition that:

- You cannot fully resolve reality.
- You do not need to.
- Partial alignment is enough.

Contrast this with existential panic. In panic, the mind demands final answers: *What happens after death? What is the ultimate meaning? Is this belief absolutely true?*

When answers are unavailable, strain escalates.

In awe, the same questions exist—but the demand for immediate resolution drops. The unknown becomes tolerated rather than attacked.

This tolerance stabilizes alignment.

Thus uncertainty feels sacred when:

- Identity relaxes.
- Narrative amplification quiets.
- Bandwidth limits are accepted.
- Incompleteness is allowed.

Spiritual experience, stripped of cultural framing, may simply be structured acceptance of partial resolution.

And this leads directly to the next distinction: Resolution seeks closure. Surrender accepts incompleteness.

We now examine that difference carefully.

15.2 — Surrender vs Resolution

The mind is conditioned to resolve. A question arises—it seeks an answer. A tension appears—it seeks closure. A contradiction surfaces—it seeks elimination.

Resolution is the natural reflex of a partial alignment system operating under strain.

But what happens when resolution is structurally impossible? When the question has no final answer? When the boundary cannot be fully articulated? When incompleteness is not temporary, but fundamental?

Here lies the distinction between resolution and surrender.

- Resolution attempts to eliminate strain.
- Surrender accepts that some strain cannot be eliminated.

Resolution is an active narrowing of possibilities until threshold collapse occurs. Surrender is relaxation of the demand for total narrowing.

This is not passivity. It is structural intelligence.

Existential Questions

Consider questions such as: *What is the ultimate meaning of life? Is there an absolute truth behind all appearances? What happens after death?*

The mind seeks definitive articulation. It wants a stable narrative anchor.

But if incompleteness is structural—if total resolution is impossible—then the attempt to force closure creates chronic alignment overload. The system remains tense.

Spiritual traditions often speak of surrender—not as defeat, but as release of rigid anchor demand.

Surrender is not abandonment of thought. It is reduction of identity rigidity. It is allowing partial alignment without demanding absolute closure.

Structural Difference

From a structural standpoint:

- Resolution narrows geometry aggressively.
- Surrender widens tolerance within bandwidth.
- Resolution seeks elimination of alternatives.
- Surrender allows alternatives to coexist without panic.

In physics, collapse exhausts viable configurations under constraint. In spirituality, surrender relaxes the compulsion to exhaust.

The difference is subtle but powerful.

When a person insists, *"I must know with certainty,"* cognitive strain intensifies. When a person says, *"I may not fully know—and that is acceptable,"* strain redistributes.

Bandwidth stabilizes. Identity softens. Presence emerges.

This is why surrender often feels relieving. It reduces alignment overload. It prevents brittle anchor fracture. It aligns the mind with structural incompleteness rather than fighting it.

Importantly, surrender is not anti-rational. It does not reject inquiry. It simply acknowledges that some boundaries cannot be fully resolved.

Resolution is appropriate for practical decisions. Surrender is appropriate for structural incompleteness.

Confusing the two produces anxiety. Demanding total resolution where only partial alignment is possible creates suffering.

Spiritual maturity may therefore lie not in possessing final answers—but in discerning where resolution is possible and where surrender is wiser.

And when surrender stabilizes bandwidth, something interesting happens: the mind becomes present within strain.

15.3 — Presence Within Strain

When resolution is impossible and surrender reduces rigidity, something subtle becomes available: presence.

Presence is often described poetically—"living in the now," "pure awareness," "mindfulness." But structurally, presence is something precise.

It is the mind observing strain without forcing immediate collapse.

Resolution Reflex vs Presence

Ordinarily, when cognitive strain arises, the system reacts:

- It amplifies narrative.
- It strengthens identity anchors.
- It searches aggressively for closure.
- It suppresses competing geometry.

This is the resolution reflex.

But when surrender softens rigidity, the mind can remain with unresolved geometry without escalating.

It does not deny strain. It does not dramatize strain. It does not rush to collapse.

It allows negotiation to continue at tolerable bandwidth.

Presence is therefore stabilized partial alignment.

Example: Anxiety

Consider anxiety.

Anxious thought loops arise when the mind tries repeatedly to force certainty where structural incompleteness persists: *What if this goes wrong? What if I made the wrong choice? What if I lose everything?*

Each attempt to resolve the uncertainty intensifies strain.

Presence interrupts the escalation.

Instead of forcing closure, attention rests on the experience of strain itself. The thought is observed. The sensation is noticed. The uncertainty is acknowledged.

But narrative amplification is reduced. Identity defensiveness softens. Bandwidth remains stable.

In this state, strain does not vanish—but it becomes manageable.

Just as classical ripple lock absorbs micro-uncertainty through density, presence absorbs cognitive uncertainty through non-reactive awareness.

The geometry shifts from aggressive narrowing to stable tolerance.

Practices of Presence

This is why contemplative practices emphasize breath, sensation, or simple observation.

They reduce narrative reinforcement. They quiet identity defense. They expand alignment span gently.

Presence does not solve existential questions. It allows coexistence with them.

It does not eliminate incompleteness. It integrates it.

And when incompleteness is integrated rather than resisted, uncertainty feels spacious instead of threatening.

Structural Difference

This is the structural difference between anxiety and spiritual stillness.

Both encounter unresolved geometry. One fights it. One allows it.

Allowing does not weaken the system. It increases resilience.

A system that tolerates strain without brittle collapse is more stable than one that demands rigid certainty.

Critical Insight

If surrender and presence stabilize partial alignment, then what destabilizes spirituality?

The same force that destabilizes physics and psychology: the illusion of absolute certainty.

We now examine how rigid certainty fractures spiritual systems just as it fractures identity and classical determinism.

15.4 — The Illusion of Absolute Certainty

If surrender stabilizes the mind and presence absorbs strain, what destabilizes spirituality? The same force that destabilizes identity: the demand for absolute certainty.

Across history, spiritual traditions began with encounters with incompleteness—vastness, mortality, awe. But over time, many evolved into systems that promised final answers: definitive truths, unquestionable doctrines, absolute interpretations.

This transformation is structurally understandable. Uncertainty produces strain. Strain seeks resolution. Narrative amplification builds stability. Identity anchors around belief. Collective reinforcement turns flexible insight into rigid certainty.

But rigidity introduces brittleness.

Just as classical determinism fails under extreme strain, spiritual certainty fails under contradiction, social change, or internal doubt.

When a system claims absolute closure, it suppresses residual strain aggressively. Suppression is not elimination. Residual geometry accumulates beneath the surface. Eventually, cracks appear:

- Internal contradictions.
- Ethical inconsistencies.
- Institutional corruption.
- Disillusionment among followers.

The more absolute the certainty, the sharper the fracture when compression arrives.

Humility vs Rigidity

The illusion lies in confusing spiritual humility with doctrinal rigidity.

- Spiritual humility recognizes incompleteness.
- Doctrinal rigidity denies it.

Humility stabilizes partial alignment. Rigidity demands total resolution.

But total resolution is structurally impossible.

Thus, systems built on absolute certainty must defend themselves constantly against emerging strain. They resist questioning, suppress doubt, and punish deviation. This defensive posture is alignment overload at collective scale.

It mirrors psychological rigidity. It mirrors classical determinism stretched beyond regime.

The irony is profound: spiritual traditions that originated from awe before incompleteness can become institutions that deny incompleteness.

When that happens, spirituality shifts from surrender to control. From presence to enforcement. From humility to certainty addiction.

Certainty addiction is the refusal to tolerate structural incompleteness. But incompleteness does not disappear. It resurfaces—through crisis, reform, schism, cultural transformation.

True spiritual stability requires recognition that no belief system can exhaust all uncertainty.

The sacred does not require final answers. It requires structural humility.

This realization leads to the final synthesis of this chapter: spiritual maturity is not possession of certainty—it is conscious alignment within incompleteness.

15.5 — Spiritual Maturity as Structural Awareness

If uncertainty is structural, if incompleteness is universal, if total resolution is impossible—then spiritual maturity cannot mean possessing final answers.

It must mean something else.

Defining Spiritual Maturity

Spiritual maturity is the conscious recognition of structural incompleteness.

It is the understanding that:

- The mind is finite.
- Bandwidth is limited.
- Identity is conditional.
- Narratives are stabilizers, not absolutes.

This recognition does not weaken belief. It refines it.

A spiritually immature system demands certainty. A spiritually mature system tolerates doubt.

Immaturity seeks permanent anchor rigidity. Maturity allows anchor flexibility.

This does not mean *anything is true.* It means no articulation exhausts all possible geometry.

Structural Awareness

Spiritual maturity understands that:

- Resolution is local.
- Surrender is intelligent.
- Presence is stabilizing.
- Certainty is conditional.

Such maturity produces humility—not submissiveness, but accurate scale awareness.

The individual recognizes themselves as a partial alignment system within a layered universe.

- The universe is not fully knowable.
- The self is not fully resolvable.
- Beliefs are tools, not final endpoints.

Social Implications

This perspective dissolves fanaticism. It reduces defensive aggression. It allows dialogue across difference.

Because when incompleteness is acknowledged, threat perception decreases. Conflict escalates when systems believe their articulation is final. Harmony increases when systems recognize partiality.

Spiritual maturity therefore aligns with structural stability.

It is not anti-scientific. It is not anti-rational. It is deeply structural.

It integrates:

- Cognitive strain,
- Narrative reinforcement,
- Identity anchoring,
- Residual uncertainty.

It recognizes that awe arises not from mystery for its own sake—but from scale beyond ego bandwidth.

The sacred is not the unknown made magical. It is the unknown accepted without collapse.

The Mature Response

This acceptance generates calm strength. It generates resilience. It generates adaptive flexibility.

The spiritually mature individual does not eliminate doubt. They integrate it.

They do not deny incompleteness. They live within it consciously.

Transition to Collective Scale

And once we see this clearly, a final shift becomes necessary.

If individuals mature through structural awareness, what happens at collective scale? Can civilizations mature? Can institutions tolerate incompleteness? Or do they amplify rigidity?

We now widen the lens once more: From individual spirituality to collective structure. From inner surrender to societal uncertainty.

From Inner Surrender to Civilizational Strain

We began this chapter with awe— with the night sky, with silence, with the feeling that the unknown is larger than us.

We now understand why. Uncertainty feels sacred when the mind stops demanding total resolution.

Surrender stabilizes where resolution cannot. Presence absorbs strain without suppression. Humility replaces rigidity. Incompleteness becomes tolerable.

Inner Stability

Spiritual maturity is not certainty. It is structural awareness.
The recognition that:

- Alignment is partial.
- Identity is conditional.
- Belief is bounded.
- Finality is illusion.

This realization produces inner stability.

Scaling the Question

But something important follows.
An individual can soften identity. An individual can tolerate doubt. An individual can practice surrender.
What about societies? What about institutions? What about nations, religions, ideologies, markets, political systems?
When millions of partial alignment systems synchronize, they produce large-scale ripple locks:

- Collective narratives.
- Collective identities.
- Collective certainties.

Aggregation amplifies stability. But aggregation also amplifies rigidity.

Collective Compression

Just as classical determinism hid micro-uncertainty, collective consensus hides residual social strain.

Until compression arrives. Until crisis exposes suppressed geometry. Until certainty fractures.

The same structure we traced in physics and psychology now reappears at civilizational scale.

The universe does not escape incompleteness. Neither does society.

The question is no longer personal. It becomes collective:

- Can systems tolerate uncertainty without collapsing into chaos?
- Can institutions mature spiritually—not religiously, but structurally?
- Can power operate without demanding absolute certainty?

We now step into the final widening of the lens: From individual alignment to collective strain. From personal surrender to societal uncertainty.

The geometry remains unchanged. Only the scale grows. And at scale, the consequences intensify.

Chapter 16 — Fear, Control, and Drift

When Uncertainty Feels Dangerous

Uncertainty can feel sacred. Under the night sky, it humbles. In silence, it softens. In contemplation, it expands.

But place the same uncertainty in a different context— a sudden loss of income, a medical diagnosis, public humiliation, political instability.

Now uncertainty does not feel sacred. It feels dangerous.

Structural Continuity

Why? The structure has not changed. Incompleteness is still present. Bandwidth is still finite. Identity is still conditional.

What changes is interpretation relative to survival.

The human nervous system evolved under physical threat conditions. Uncertainty in the wild meant potential death. Ambiguous rustling in the bushes could signal a predator. Delay in reaction reduced survival probability.

Thus, the brain became tuned to treat certain uncertainties as urgent survival deltas.

A survival delta is strain interpreted as threat to persistence.

When such strain is detected, bandwidth compresses rapidly. Attention narrows. Decision speed increases. Tolerance for ambiguity decreases.

Fear is fast alignment under perceived survival instability. It is not irrational. It is accelerated threshold logic.

The system attempts to collapse competing possibilities quickly to restore stability.

Symbolic Threats

In modern life, however, many threats are symbolic rather than physical:

- Reputation loss.
- Ideological disagreement.
- Status decline.
- Social rejection.

Yet the nervous system often responds as if survival itself were at stake.

The same compression occurs:

- Bandwidth narrows.
- Narrative amplifies.
- Identity hardens.
- Control reflex activates.

Uncertainty becomes intolerable. Fear transforms incompleteness from spacious to hostile.

This reveals something essential: uncertainty itself is neutral.

It becomes sacred when identity relaxes. It becomes dangerous when survival coding activates.

The difference lies not in reality—but in interpretation of strain relative to persistence.

The mind cannot rest because incompleteness persists. But it panics when incompleteness is perceived as existential threat.

This chapter explores how fear arises structurally, how control becomes a compensatory illusion, and how suppressed strain produces drift.

We begin with the most fundamental layer: survival delta.

16.1 — Survival Delta

Fear is not philosophical. It is biological.

Long before humans contemplated infinity or spiritual surrender, the nervous system evolved to detect threat. In evolutionary terms, uncertainty was rarely abstract. It was immediate:

- Is that sound a predator?
- Is that stranger hostile?
- Is this food safe?

Ambiguity meant risk.

The brain therefore developed rapid pattern-recognition mechanisms that treat certain uncertainties as urgent deltas requiring fast resolution.

This is survival delta.

Structural Shifts

Survival delta is strain interpreted as threat to persistence.

When such strain is detected, several structural shifts occur almost instantly:

- Bandwidth narrows.
- Attention sharpens toward threat cues.
- Competing narratives are suppressed.
- Decision thresholds lower.

- Reaction speed increases.

This compression is adaptive. Under physical danger, prolonged negotiation reduces survival probability. The system must collapse possibilities quickly—even if imperfectly.

Fear is therefore accelerated alignment. It is the mind forcing rapid threshold collapse under compressed bandwidth.

Evolutionary Bias

Fear is not irrational. It is structural prioritization.

The organism temporarily sacrifices complexity for speed.

In the wild, this worked effectively. If a rustling bush was a predator, immediate flight increased survival odds. If it was wind, the false positive cost was small.

Evolution favored sensitivity. Better to overreact than underreact.

This bias remains embedded in modern humans.

Modern Mismatch

However, modern environments differ from ancestral conditions. Most contemporary threats are not immediate physical predators. They are social, symbolic, or economic.

Yet the survival circuitry does not distinguish sharply between physical and symbolic threat.

- A job loss can trigger similar compression as physical danger.
- Public criticism can activate identity threat as if survival were at risk.
- Ideological disagreement can be processed as group

exclusion—a historically dangerous scenario.

When symbolic strain activates survival coding, bandwidth collapses. Nuance disappears. Ambiguity becomes intolerable. Control reflex strengthens.

This is why fear narrows perspective.

It is not moral weakness. It is bandwidth compression under perceived persistence instability.

The Cost

However, there is a cost.

Repeated survival activation in non-lethal contexts creates chronic alignment overload. The system remains tense. Narrative amplification intensifies. Identity hardens defensively. Residual strain accumulates beneath surface certainty.

Modern humans live in symbolic environments that continuously trigger survival delta without offering physical resolution.

This mismatch generates anxiety cultures.

Structural Insight

Understanding survival delta structurally helps reduce shame around fear.

Fear is not evidence of failure. It is evidence of compressed bandwidth.

The problem is not fear itself. The problem is when symbolic amplification hijacks biological circuitry.

We now examine how modern symbolic systems magnify survival delta far beyond its evolutionary design.

16.2 — Symbolic Amplification

If survival delta were triggered only by physical danger, modern life would be far calmer. But humans do not live in purely physical environments. We live in symbolic worlds:

- Language.
- Money.
- Reputation.
- Status.
- Belief systems.
- Political identity.

These are not predators in the bushes. Yet they activate survival circuitry.

Why Symbolic Threats Matter

The brain encodes social belonging and status as survival-relevant. For most of human history, exclusion from the group meant vulnerability. Loss of status meant reduced access to resources. Social rejection carried real risk.

Thus, symbolic threat became biologically linked to persistence.

In modern environments, symbolic systems have scaled massively. A disagreement online. A shift in political power. A rumor about reputation. A change in market conditions. Each can trigger survival delta.

The nervous system compresses bandwidth just as it would under physical threat.

This is symbolic amplification: abstract uncertainties interpreted as existential risks.

Structural Pattern

The structure unfolds predictably:

- A narrative frames the event as dangerous.
- Identity anchors around the narrative.
- Social reinforcement synchronizes response.
- Survival circuitry activates.
- Fear escalates.

The threat may not involve physical harm. But the body responds as if it does.

This is why ideological conflict feels intense. It is rarely about facts alone. It is about identity-coded survival.

When belief becomes identity, disagreement becomes perceived threat. When identity becomes tribal, opposition becomes existential.

Bandwidth narrows. Nuance disappears. Certainty hardens.

Media and Chronic Compression

Symbolic amplification also explains media-driven anxiety. Continuous exposure to threat narratives—economic collapse, cultural decline, technological disruption—maintains survival delta in a chronic state.

The nervous system remains in low-grade compression.

Over time, chronic compression produces rigidity. People seek control. They seek certainty. They seek strong narratives that promise stability.

This prepares the ground for the next structural move: the control illusion.

16.3 — The Control Illusion

When survival delta activates, the system demands stability. Bandwidth has narrowed. Threat has been detected. Uncertainty feels dangerous.

The mind now seeks rapid resolution. And the most immediate psychological strategy available is control.

Control promises restoration of alignment:

- *"If I manage this precisely, nothing will go wrong."*
- *"If we enforce strict rules, instability will disappear."*
- *"If we eliminate opposing views, certainty will return."*

Control feels stabilizing because it compresses geometry artificially. It reduces visible alternatives. It narrows possibility space. It simulates total alignment.

Structural Problem

But here lies the structural problem: control suppresses strain; it does not eliminate it.

Just as rigid classical determinism suppresses quantum uncertainty at scale, rigid psychological control suppresses cognitive uncertainty temporarily.

The system appears stable. But residual geometry persists beneath enforcement.

Personal Scale

Consider personal life. A person experiencing anxiety may attempt to control every detail—schedule, relationships, environment. For a time, structure reduces visible unpredictability.

But the world does not fully submit to control. Unexpected events reintroduce strain. The more rigid the control, the sharper the fracture when unpredictability returns.

Control becomes brittle alignment.

Social Scale

At collective scale, the pattern magnifies. Under collective fear:

- Institutions centralize authority.
- Ideologies demand conformity.
- Dissent is suppressed.
- Certainty is institutionalized.

This produces short-term stability. But suppressed strain accumulates.

Alternative narratives go underground. Misalignment grows invisibly. When pressure exceeds reinforcement capacity, sudden rupture occurs: revolutions, institutional collapse, cultural fragmentation.

The illusion of total control collapses dramatically.

Misapplied Resolution Reflex

Structurally, control is an attempt to force total resolution within a system that cannot achieve total resolution. It is misapplied resolution reflex in a fundamentally incomplete universe.

The deeper the fear, the stronger the control impulse. The stronger the control, the greater the suppressed strain.

This explains why rigid systems often fail catastrophically rather than gradually. Flexible systems absorb strain. Rigid systems accumulate it.

Drift

Control also distorts perception. When individuals believe they are fully in control, they underestimate uncertainty. They interpret suppressed strain as eliminated risk.

Drift begins.

Decisions become detached from structural reality. Overconfidence grows.

This prepares the next structural consequence: drift under suppression.

We now turn to that slow and dangerous process.

16.4 — Drift Under Suppression

When fear activates survival delta and control attempts to compress uncertainty, the system may appear stable. But stability under suppression is not equilibrium. It is tension held artificially.

Over time, suppressed strain does not disappear. It drifts.

Drift is slow misalignment between perceived stability and structural reality.

Unlike collapse, drift is quiet. There are no dramatic threshold events at first.

Instead:

- Small inconsistencies are ignored.
- Minor contradictions are rationalized.
- Feedback signals are dismissed.
- Alternative narratives are silenced.

Control maintains surface order. But underneath, geometry shifts.

Examples of Drift

In organizations: A company becomes overly confident. Leadership suppresses dissent. Data that contradicts the dominant narrative is minimized. Short-term stability persists. Yet market conditions change. Innovation lags. Internal morale weakens. No single event signals collapse. But misalignment accumulates. Drift continues. Eventually, a shock reveals accumulated strain. What appears to be sudden failure is actually slow drift reaching threshold.

In personal life: An individual suppresses doubt about career dissatisfaction. Identity narrative reinforces "success." Control strategies maintain routine. Years later, burnout appears "suddenly." But burnout is threshold collapse after prolonged drift.

At societal scale: Collective fear justifies rigid policies. Control suppresses visible dissent. Stability appears strong. But suppressed social strain—economic inequality, cultural fragmentation, generational discontent—accumulates. Because uncertainty is denied rather than integrated, flexibility declines. When compression exceeds tolerance, crisis erupts. The system blames external shock. But the root was drift.

Structural Truth

Drift occurs because suppressed strain is not integrated into alignment recalibration. Suppression narrows bandwidth artificially. It prevents early adaptation. The system loses sensitivity to weak signals.

Over time, reality and narrative diverge. The greater the divergence, the sharper the eventual correction.

Drift is therefore not chaos. It is misalignment under chronic suppression.

It reveals a critical structural truth: control without awareness destabilizes over time.

To prevent drift, the system must integrate uncertainty consciously rather than eliminate it.

This brings us to the final stabilizing principle of this chapter: awareness anchoring.

16.5 — Awareness Anchoring

If fear compresses bandwidth and control suppresses strain, what stabilizes without brittleness? Awareness.

Not passive observation. Not spiritual abstraction. Structural awareness.

Definition

Awareness anchoring means recognizing strain without immediately forcing collapse. It is the capacity to detect survival delta without over-amplifying it.

When fear arises, awareness asks:

- Is this physical threat? Or symbolic threat?
- Is this immediate danger? Or narrative amplification?

That pause widens bandwidth. It prevents automatic compression.

Unlike rigid control, awareness does not deny uncertainty. It acknowledges incompleteness consciously.

Structural Function

This shifts the geometry. Instead of narrowing possibilities prematurely, awareness allows competing geometries to coexist briefly within tolerable span.

This reduces drift. Because drift emerges from suppression, and awareness prevents suppression.

Examples

Leadership: A leader acknowledges uncertainty openly. Rather than projecting false certainty, they admit incomplete information. This reduces symbolic amplification. It prevents overconfidence. It keeps alignment adaptive.

Personal scale: When doubt arises, instead of suppressing it or panicking, the individual observes: *There is strain here.* They neither escalate narrative nor deny tension. They integrate it gradually. This prevents survival delta from hijacking identity.

Flexible Anchoring

Awareness anchoring is flexible. It adjusts under new information. It recalibrates alignment continuously.

Rigid control says: *"I must eliminate uncertainty."* Awareness says: *"Uncertainty is structural. I will navigate within it."*

The difference determines long-term stability.

Flexible anchoring absorbs strain incrementally. Rigid anchoring fractures under compression.

Collective Scale

Awareness also interrupts symbolic amplification. Modern environments constantly generate artificial survival triggers—media alerts, ideological conflict, economic volatility.

Without awareness, the nervous system remains in chronic compression. With awareness, the individual can differentiate signal from amplification.

Bandwidth remains wider. Alignment remains adaptive.

This is not denial of risk. It is calibrated response.

In structural terms: Awareness maintains partial alignment without suppressing residual strain. It accepts incompleteness without surrendering functionality. It is dynamic stability.

And when awareness scales collectively—through transparent institutions, open dialogue, tolerance of dissent—societies reduce drift and brittleness.

But when fear scales without awareness, control intensifies, drift accumulates, and collective instability grows.

From Personal Fear to Collective Instability

We began this chapter with a simple contrast: uncertainty can feel sacred, or uncertainty can feel dangerous.

The difference was not in reality. It was in interpretation.

When incompleteness is perceived as threat to survival, bandwidth compresses. Fear activates. Control intensifies. Suppression begins. And suppression creates drift.

We have traced the sequence clearly:

- Survival delta narrows attention.
- Symbolic amplification escalates threat perception.
- Control attempts simulate stability.
- Suppressed strain accumulates.

- Drift separates narrative from reality.
- Threshold collapse follows.

This pattern is not psychological accident. It is structural.

The same geometry we observed in physics—compression, overload, collapse—reappears in human systems.

Fear is accelerated alignment under perceived instability. Control is forced resolution under incomplete geometry. Drift is slow misalignment under suppression. Awareness anchoring is adaptive partial alignment.

At individual scale, this determines resilience. But scale changes consequences.

When one mind compresses under fear, a life destabilizes. When millions compress simultaneously, civilizations destabilize.

Collective fear amplifies faster than individual fear. Social media synchronizes survival delta across populations. Ideologies transform symbolic disagreement into existential threat. Institutions promise certainty to calm anxiety. Control expands. Suppression deepens. Drift accelerates.

History is full of such cycles: economic panics, cultural purges, authoritarian consolidations, revolutions.

These are not random explosions. They are large-scale alignment overload under survival coding.

The geometry remains the same. Only the magnitude increases.

The final movement of this book must therefore examine collective uncertainty directly: how societies stabilize, how they drift, how they collapse, how they mature.

If incompleteness is structural, then no system—political, economic, religious—can eliminate uncertainty.

The question is not whether uncertainty exists. The question is how systems respond to it.

With fear and control? Or with awareness and adaptive alignment?

We now step into the collective field. From individual survival delta to systemic instability. From private drift to historical tipping points.

The geometry is consistent. The stakes are larger.

Interlude — When Uncertainty Enters the Human Heart

Until now, uncertainty lived in particles, stars, collapse, and curvature. In this part, it entered the human field.

We discovered something profound: The structure did not change. Only the scale did.

Core Realizations

Psychological Layer

- The mind is a partial alignment system, not a certainty machine.
- Cognitive strain is the mental equivalent of delta.
- Decisions are local threshold collapses, not total resolution.
- Residual doubt persists because alignment is always partial.
- Narratives act as psychological aggregation, suppressing visible uncertainty.
- Identity functions as a geometric anchor, stabilizing cognitive strain.
- Existential doubt arises when anchor geometry is threatened.
- Psychological maturity is the ability to tolerate incompleteness consciously.

Spiritual Layer

- Uncertainty feels sacred when resistance drops.
- Awe is scale-awareness without defensive contraction.
- Surrender is not defeat—it is acceptance of structural incompleteness.
- Presence stabilizes bandwidth without forcing collapse.
- Absolute certainty is illusion; rigidity produces brittleness.
- Spiritual maturity is structural humility, not possession of final answers.

Fear and Control Layer

- Fear is survival delta under perceived instability.
- Modern symbolic systems amplify biological threat circuitry.
- Control is forced resolution under compressed bandwidth.
- Suppression of strain leads to drift.
- Drift accumulates misalignment beneath surface stability.
- Awareness anchoring prevents brittleness and reduces amplification.

Unified Insight

The geometry of uncertainty is consistent across:

- Physics
- Psychology
- Spirituality
- Survival behavior

Incompleteness is not a flaw—it is continuity. Certainty is conditional. Stability requires adaptive partial alignment. Collapse occurs when compression exceeds tolerance. Growth occurs when incompleteness is integrated.

Closing

This part closes with clarity: Uncertainty is not just an external feature of reality. It is the structural condition of consciousness itself.

And once consciousness is understood as incomplete, the final question emerges naturally:

What happens when incomplete minds build collective systems? When identities synchronize? When narratives scale? When fear aggregates?

The next movement explores uncertainty at civilizational scale. From the human mind to the human world.

The geometry remains unchanged. The consequences expand.

Part VI — Civilization Under Strain

"Civilizations do not collapse because uncertainty exists. They collapse because they refuse to align with it."

Overview

Until now, uncertainty has moved across scales:

From quantum strain to classical aggregation, from black holes to identity anchors, from fear to spiritual humility.

Now the lens widens one final time.

What happens when partial alignment systems—human minds—organize into collective structures?

Civilizations are not static entities. They are large-scale ripple locks built from synchronized narratives, identities, and survival deltas.

This part examines how societies manage uncertainty—or fail to.

- **Ideologies** act as delta management systems. They compress ambiguity into stable narratives. When rigid, they accumulate suppressed strain. When adaptive, they recalibrate and survive.
- **Technology** introduces artificial resolution. Machines collapse possibilities without experiencing uncertainty. They generate coherence without awareness. They force articulation at scale—sometimes faster than human bandwidth can integrate.
- **Progress** is not linear certainty. It is threshold shift after

threshold shift. Knowledge expands, awareness evolves, and equilibrium becomes a moving frontier.

Civilizations endure not by eliminating uncertainty—but by integrating it adaptively.

When suppression dominates, drift accumulates. When awareness scales, longevity increases.

This part completes the arc of this book:

Uncertainty is not chaos.

It is the structural engine of continuity— in matter, in mind, and in civilization.

The question is no longer whether uncertainty exists.

The question is whether we can align with it—collectively.

Chapter 17 — Ideologies as Delta Management

Civilization as Structured Strain

Civilization is often described as order. But order is not the absence of tension. It is tension arranged.

Every society lives inside competing demands. Security competes with freedom. Tradition competes with innovation. Identity competes with universality. Growth competes with sustainability.

These are not temporary disagreements. They are structural deltas.

At the level of the individual, uncertainty arises when competing distortions cannot fully resolve within finite alignment bandwidth. At the level of civilization, the same condition persists — scaled, distributed, and institutionalized.

A society is not a unified mind. It is a field of interacting minds. Each carrying its own strain geometry. Each attempting articulation.

What we call stability is not consensus. It is managed incompletion.

Ideologies emerge within this condition. Not as accidents. Not merely as belief systems. But as alignment architectures.

An ideology is a geometry built to manage persistent distortion within a population.

It offers resolution pathways. It prioritizes certain deltas over others. It reduces cognitive overload by simplifying articulation. It converts complexity into narrative.

In doing so, it stabilizes collective action.

But simplification is never neutral. Every resolution suppresses alternative strain. Every alignment excludes competing articulation.

Civilization, therefore, is not the elimination of uncertainty. It is the organization of it.

The question is never whether strain exists. It is how it is distributed.

When strain is circulated, societies appear dynamic but resilient. When strain is suppressed, they appear stable but brittle.

This is not a moral claim. It is structural.

Just as atoms persist through ripple lock rather than absolute stillness, societies persist through managed tension rather than complete agreement.

Uncertainty at civilizational scale is not chaos. It is unresolved geometry seeking articulation.

Ideology is one of the primary mechanisms through which that articulation is attempted.

17.1 — What an Ideology Actually Is

An ideology is often described as a set of beliefs. But beliefs are only the visible surface.

Structurally, an ideology is a coordination mechanism.

When many minds coexist within the same civil field, each carries its own deltas — preferences, fears, aspirations, memories, expectations. These do not naturally align. Left unmanaged, they produce friction, fragmentation, and directional drift.

An ideology reduces that drift.

It does so by providing:

A shared narrative. A shared hierarchy of values. A shared definition of threat and purpose. A shared interpretation of uncertainty.

In structural terms, it narrows alignment pathways.

Where infinite articulations are possible, ideology selects a subset and declares them primary. This reduces bandwidth demand. It simplifies decision-making. It converts ambiguity into direction.

This is not inherently manipulative. It is functional.

Just as in physics, when multiple competing deltas exist, partial stabilization requires prioritization. A system cannot resolve all distortions simultaneously. It selects alignment channels.

Societies do the same.

An ideology tells a population:

What matters most. What can be deferred. What is acceptable strain. What is intolerable strain.

In doing so, it compresses complexity into manageable form.

Without such compression, collective action becomes incoherent. Every decision would require full articulation of all competing demands. Alignment would stall. The system would remain in continuous deliberative overload.

Ideology therefore acts as a macro-scale resolution shortcut.

It does not eliminate uncertainty. It manages its distribution.

But this management comes at a cost.

When alignment pathways narrow, certain deltas are amplified while others are suppressed. Some narratives gain articulation; others remain residual. Some identities feel represented; others feel compressed.

Residual strain does not vanish. It accumulates.

This is why ideology must be understood structurally, not morally.

It is neither inherently virtuous nor inherently destructive. It is an organizing geometry imposed on distributed uncertainty.

At the individual level, belief reduces cognitive strain by offering explanatory closure. At the collective level, ideology performs the same function — it offers civilizational closure.

Closure creates direction. Direction enables movement. Movement sustains continuity.

But closure is always partial.

A society that forgets this begins to mistake simplification for truth and prioritization for permanence. When that happens, ideology shifts from management tool to rigidity anchor.

Understanding ideology as delta management clarifies its role. It is not the enemy of uncertainty. It is one method of containing it.

The question is not whether societies will generate ideologies. They must.

The question is how flexible those alignment geometries remain when new strain emerges.

17.2 — Rigid vs Adaptive Systems

Not all ideologies manage strain in the same way. Some compress it. Others circulate it.

The difference is not moral. It is structural.

A rigid ideological system seeks rapid alignment. It minimizes visible disagreement. It narrows articulation channels. Competing deltas are treated as threats rather than signals. Uncertainty is interpreted as weakness.

Such systems often appear strong.

They project clarity. They reduce deliberation. They produce decisive movement.

From within, this feels stable. Strain appears resolved because alternative articulation is suppressed. But suppression is not resolution. It is deferred alignment.

Residual deltas remain embedded within the civil field. They are not erased; they are unexpressed.

When articulation bandwidth is restricted, unresolved strain accumulates beneath the surface. Over time, the system's visible coherence masks internal density. The more certainty is demanded, the less redistribution occurs. Alignment becomes forced rather than negotiated.

Forced alignment reduces short-term instability. It increases long-term fragility.

Adaptive ideological systems operate differently.

They tolerate competing articulation. They permit critique. They allow narratives to evolve. This does not eliminate strain. In fact, it often increases visible tension. Disagreement becomes public. Alignment appears slower. Stability looks uncertain.

But the redistribution is continuous.

Competing deltas are given articulation pathways. Feedback loops remain active. Alignment is recalibrated rather than imposed. The system absorbs distortion incrementally instead of compressing it into silence.

Adaptive systems appear noisy. Rigid systems appear calm.

The calm of rigidity is low-visibility strain. The noise of adaptability is managed redistribution.

This distinction mirrors earlier structural principles.

In physical systems, dynamic equilibrium requires continuous micro-adjustment. Absolute stillness is not stability; it is frozen tension. Similarly, civil systems that allow micro-articulation maintain dynamic equilibrium. Those that eliminate it shift toward brittle coherence.

Rigidity reduces uncertainty by narrowing bandwidth. Adaptability reduces uncertainty by expanding bandwidth.

Both aim at stabilization. Only one preserves resilience.

No ideology is purely rigid or purely adaptive. All systems contain elements of both. The structural question is proportional: how much articulation is permitted before alignment is enforced?

When articulation capacity declines, rigidity increases. When feedback channels narrow, strain concentrates. When concentration exceeds civilizational ATT — the system's alignment threshold — instability becomes abrupt rather than gradual.

Adaptive systems rarely avoid strain. They metabolize it.

Rigid systems rarely display strain. They store it.

Understanding this distinction clarifies why some societies endure turbulence without collapse, while others fracture under pressures that seemed invisible just moments before.

The issue is not ideology itself. It is the geometry of its flexibility.

17.3 — Accumulated Unresolved Strain

Strain that is not articulated does not disappear. It redistributes. It densifies.

In a civil field, unresolved deltas accumulate when articulation bandwidth narrows. Individuals may adapt privately. Groups may suppress disagreement publicly. Institutions may delay response. But structural tension continues to exist beneath visible order.

This accumulation is gradual.

At first, it produces minor distortions — small inconsistencies between narrative and lived experience. Policies that no longer align with reality. Promises that no longer stabilize expectation. Language that feels increasingly disconnected from circumstance.

These are not dramatic events. They are micro-misalignments.

If articulation channels remain open, these distortions are processed incrementally. Feedback enters the system. Adjustment occurs. Residual strain redistributes.

If articulation channels narrow, the distortions persist.

When dissent is framed as disruption rather than signal, when reinterpretation is labeled instability, when uncertainty is equated with weakness, the system begins to treat strain as illegitimate.

But geometry does not respond to labels.

Competing deltas continue to exert alignment demand.

Over time, the density of unresolved strain increases. This density is not always visible in economic indicators, institutional performance, or public ritual. Outward coherence may even intensify. The more strain accumulates, the more forcefully narrative unity is projected.

This projection is an attempt at stabilization.

Yet stabilization without redistribution increases internal pressure.

Civilizational ATT — its alignment threshold time — is finite. Every society has limited bandwidth for absorbing distortion before reconfiguration becomes necessary. When competing demands exceed this bandwidth, delayed articulation converts into abrupt transition.

What appears sudden is rarely spontaneous. It is threshold crossing.

At that point, long-suppressed deltas seek rapid articulation. Systems that once seemed immovable shift quickly. Narratives fragment. Alliances reorganize. Institutions struggle to recalibrate.

Observers often interpret such events as moral collapse, corruption, or collective irrationality. Sometimes those elements are present. But structurally, the deeper cause is alignment overload.

Uncertainty that was denied returns amplified.

The accumulation phase is quiet. The release phase is visible.

Understanding this dynamic reframes civil instability. It is not always the result of malicious actors or intellectual failure. Often it is the consequence of unresolved strain exceeding alignment capacity.

When a system cannot metabolize uncertainty gradually, it experiences it abruptly.

This is not a prediction of collapse. It is a description of threshold behavior.

Just as in physical systems, where compressed distortion eventually reconfigures the field, civil fields undergo reorganization when accumulated deltas can no longer remain unarticulated.

The question is not whether strain will accumulate. It always does.

The question is whether a society maintains sufficient alignment bandwidth to process it before threshold is reached.

17.4 — Collapse Dynamics

Collapse is often imagined as breakdown. But structurally, it is reconfiguration under overload.

When accumulated strain exceeds a system's alignment bandwidth, articulation no longer proceeds incrementally. The geometry shifts abruptly. Institutions that once stabilized distortion fail to absorb competing deltas. Narratives that once synchronized perception lose coherence.

This does not always begin with visible chaos. It may begin with subtle decoupling.

Policies stop producing expected outcomes. Institutions continue functioning but lose legitimacy. Language remains familiar but no longer persuades.

These are early indicators of alignment fatigue.

As misalignment grows, feedback loops degrade. Signals from the periphery do not reach the center in time. Decision cycles slow. Corrections become reactive rather than anticipatory. The system begins operating on outdated internal maps.

At this stage, rigidity often increases. Certainty is amplified to compensate for internal instability. Simplification becomes more aggressive. Competing articulation is treated as destabilizing rather than diagnostic.

But compression accelerates overload.

When articulation pathways are too narrow, strain redistributes along uncontrolled channels. It may appear as sudden protest, institutional rupture, fragmentation of authority, or rapid ideological replacement. The specific form varies. The structural mechanism remains consistent: accumulated deltas seek release.

Collapse, therefore, is not the disappearance of order. It is the rapid reorganization of alignment geometry.

What collapses is not reality. It is a particular configuration of coherence.

In physical systems, threshold crossing produces phase transition. The material does not vanish; it rearranges. Similarly, civil systems rarely dissolve into nothingness. They reorganize around new alignment priorities.

The perception of collapse is heightened because the shift occurs faster than the system's previous equilibrium cycle. What was once stable becomes unstable in compressed time. The contrast produces shock.

But shock is relative to expectation.

If uncertainty had been acknowledged earlier, the transition would appear less catastrophic. When uncertainty is denied, its eventual articulation feels violent.

This is why suppression creates brittleness. Not because rigidity is immoral. But because delayed redistribution intensifies reconfiguration.

Collapse dynamics are not proof of failure in absolute terms. They indicate that a previous alignment geometry reached its limit.

Every civilization operates under finite ATT. Every ideology has threshold conditions.

When bandwidth is exceeded, change is no longer gradual.

Understanding collapse structurally removes unnecessary mystification. It is not always decay. It is often overdue recalibration occurring under compressed articulation.

The more a system resists micro-adjustment, the more dramatic macro-adjustment becomes.

Uncertainty, when denied, does not disappear. It reorganizes the field.

17.5 — Alignment Longevity

If collapse is threshold overload, then longevity is threshold management.

No civil system eliminates strain. It extends its ability to process it.

Longevity is not the absence of crisis. It is the capacity to metabolize competing deltas before they densify beyond alignment bandwidth. A durable society is not one that avoids uncertainty. It is one that distributes it continuously.

Distribution requires articulation.

When multiple perspectives are allowed structured expression, strain disperses across the field rather than concentrating at hidden nodes. Disagreement becomes part of system function instead of an external threat. Feedback becomes diagnostic rather than destabilizing.

This does not guarantee harmony. It guarantees recalibration.

Alignment longevity depends on three structural properties.

First, permeability of narrative. A system must allow reinterpretation without perceiving it as collapse. When narrative is flexible, new strain integrates gradually.

Second, redundancy of articulation channels. Multiple pathways for feedback prevent bottleneck accumulation. If one channel narrows, others compensate.

Third, tolerance for ambiguity. Not every distortion must be resolved immediately. Some deltas can remain partially articulated without threatening coherence.

These properties expand effective civil ATT. They increase the system's alignment bandwidth.

Rigid systems shorten their own threshold window by insisting on premature closure. Adaptive systems extend theirs by allowing incomplete articulation to circulate.

Longevity is therefore not ideological purity. It is geometric flexibility.

A civilization survives not because it is certain, but because it recalibrates. It absorbs misalignment without demanding total resolution. It recognizes that stability is sustained tension, not uniformity.

The most enduring systems do not suppress uncertainty. They normalize it.

In doing so, they convert uncertainty from destabilizing force into adaptive signal.

Alignment longevity is quiet. It does not announce itself as strength. It manifests as continuity — the ability to remain coherent while changing configuration.

Civilizations that understand this do not chase absolute certainty. They maintain dynamic equilibrium.

And dynamic equilibrium, as established earlier, is not stillness.

It is tension held gracefully.

The Structural Lesson

When viewed structurally, ideology is neither villain nor savior. It is mechanism.

It exists because distributed minds require coordination. It persists because collective alignment cannot occur without simplification. But like all mechanisms that manage distortion, its effectiveness depends on how it handles residual strain.

The structural lesson is not that ideology must be eliminated. That would be impossible. The lesson is that alignment systems must remain aware of their own limits.

Every ideology simplifies. Every simplification excludes. Every exclusion produces residual delta.

If a system forgets this, it begins to mistake temporary alignment for permanent resolution. Certainty becomes institutionalized. Reinterpretation becomes resistance. Feedback becomes threat.

At that point, uncertainty is no longer treated as structural condition. It is treated as anomaly.

But uncertainty is not anomaly. It is inevitability.

A civil system that recognizes uncertainty as fundamental does not seek to abolish it. It designs around it. It builds institutions that expect strain. It constructs processes that redistribute tension rather than compress it. It accepts partial articulation as normal.

This acceptance does not weaken coherence. It stabilizes it.

The most fragile systems are those that promise final answers. Absolute closure generates short-term cohesion but long-term brittleness. When certainty is marketed as permanence, every new distortion feels like betrayal.

In contrast, systems that normalize incompletion cultivate resilience. They understand that alignment is provisional. That recalibration is ongoing. That stability is managed tension, not finished agreement.

This does not eliminate disagreement. It integrates it.

The structural lesson of civilizational strain mirrors the ontological foundation established earlier. Just as nothing in the physical field is absolutely still, no social field is absolutely resolved. Persistence emerges from dynamic equilibrium.

A society that cannot tolerate uncertainty will eventually experience it as rupture.

A society that integrates uncertainty experiences it as evolution.

The difference lies not in ideology itself, but in the geometry of its flexibility — in whether it circulates strain or stores it.

In the end, uncertainty is not a threat to civilization. It is the condition under which civilization must operate.

Systems that understand this endure.

Chapter 18 — Technology and Artificial Resolution

Completion Without Experience

Human beings live inside uncertainty.

We hesitate. We doubt. We revise. We suspend judgment.

These are not weaknesses. They are expressions of finite alignment bandwidth. Competing cognitive deltas cannot always resolve immediately. Articulation requires time. Sometimes it remains incomplete.

Technology operates differently.

Modern artificial systems are designed to reduce ambiguity. They answer questions, classify inputs, optimize outcomes, and predict trajectories. Where uncertainty appears, they are tasked with resolution.

But resolution in a machine is not the same as resolution in a mind.

A human being experiences strain before articulation. There is internal density — tension between alternatives, concern about consequence, awareness of limitation. Uncertainty is felt as incompletion.

A machine does not feel incompletion.

It processes according to defined parameters. It evaluates probability distributions. It selects outputs that maximize coherence relative to its objective function. Where ambiguity exists, it calculates. Where multiple responses are possible, it ranks.

It completes.

Completion in artificial systems is structural requirement, not existential negotiation. Silence is treated as failure. Indecision is treated as inefficiency. Output is expected.

This creates a subtle distinction.

Technology appears to eliminate uncertainty because it produces answers. But producing an answer is not the same as resolving strain.

Artificial systems do not hold pre-resolved tension. They do not experience competing identity anchors. They do not navigate survival deltas. They execute forced articulation within programmed thresholds.

In earlier chapters, uncertainty was framed as persistent distortion within finite alignment bandwidth. Human articulation emerges from that condition. Civilization organizes around it. Ideology attempts to manage it.

Technology now participates in that process.

It does not abolish uncertainty. It changes how and where articulation occurs.

The question, then, is not whether machines can reduce ambiguity. They can.

The deeper question is what kind of reduction is taking place — and what remains unresolved beneath automated completion.

Understanding this distinction requires structural clarity.

Completion can be engineered. Experience cannot.

And the difference between the two shapes how uncertainty evolves in technological civilization.

18.1 — Machines Do Not Experience Strain

The distinction must be precise.

A human mind operates within competing internal deltas. Desire conflicts with caution. Identity conflicts with adaptation. Memory interferes with expectation. Every decision carries density because alignment affects survival, belonging, continuity.

Uncertainty, in this context, is lived.

It is not merely a probability distribution. It is tension between incompatible articulations within finite cognitive bandwidth. The delay before decision is not computational lag; it is experiential processing.

Machines do not operate within this structure.

An artificial system receives input. It maps that input through learned parameters. It produces output according to optimization targets. Its uncertainty is represented numerically — confidence scores, probability vectors, loss gradients.

But numerical uncertainty is not strain.

A machine does not hesitate because it feels the weight of alternatives. It calculates because its architecture requires resolution. There is no internal survival delta anchoring its articulation. No existential consequence attached to output. No identity continuity threatened by error.

This does not make machines trivial. It clarifies their ontology.

Artificial systems simulate selection under ambiguity. They do not endure ambiguity.

When a human mind suspends judgment, unresolved strain remains present in awareness. The incompletion is experienced. When a machine produces no output, it is not experiencing incompletion. It is either still processing or encountering a constraint within its defined parameters.

The structural difference lies in anchoring.

Human articulation is anchored in continuity of self — memory, identity, and survival geometry. Competing deltas press against that anchor. Alignment has consequence.

Machine articulation is anchored in objective functions. Competing outputs are ranked relative to performance metrics. Alignment has no internal meaning beyond optimization.

Thus, when we say machines manage uncertainty, we must specify the layer.

They manage informational distribution. They do not experience ontological incompletion.

Their architecture contains no pre-resolved strain waiting for articulation. It contains weighted connections adjusting under data exposure. When they generate output, they are not relieving tension. They are executing mapping.

This distinction matters because technological systems are increasingly embedded in civil articulation loops. When their output is treated as resolution, it can create the illusion that uncertainty itself has diminished.

But what has diminished is visible ambiguity, not structural incompletion.

Machines do not carry strain. They process representations of it.

Understanding this prevents both exaggeration and dismissal. Artificial systems are powerful tools for managing complexity. Yet their power arises from computation, not experience.

Uncertainty remains fundamental because the field in which both humans and machines operate remains governed by finite alignment thresholds.

The machine resolves input. The human navigates strain.

The two processes intersect, but they are not identical.

18.2 — Completion vs Awareness

The difference between human and artificial articulation becomes clearer when we examine how each handles incompletion.

A human mind can remain suspended.

It can say, *I do not know.* It can wait. It can defer judgment. It can tolerate ambiguity without immediate closure.

This suspension is not inactivity. It is active containment of competing deltas within finite cognitive bandwidth. The strain remains present. Awareness holds it without forcing premature articulation.

Artificial systems are not built to suspend in this way.

They are optimized to produce output.

When presented with input, they calculate and complete. Silence is interpreted as system failure. Non-response is treated as malfunction. In most architectures, the goal is not contemplative pause but resolution within performance constraints.

Completion is therefore structural requirement.

The system must return an answer, even when the underlying representation contains uncertainty. Probabilities may vary. Confidence scores may be low. But output is delivered.

This creates a fundamental divergence.

Human articulation emerges from tension. Artificial articulation bypasses tension.

In the human case, delay is integral to experience. It is part of alignment. In the artificial case, delay is computational cost to be minimized.

The human mind can resist closure when competing deltas are too dense. It can accept unresolved strain as condition of reality. Artificial systems cannot existentially accept incompletion. They are designed to converge.

This convergence is not awareness.

Awareness includes self-referential anchoring — the capacity to experience uncertainty as internal state. Artificial systems operate without such anchoring. They evaluate representations but do not inhabit them.

Completion, therefore, should not be confused with understanding.

A machine may produce a coherent response that integrates multiple variables. It may approximate human articulation with remarkable fluency. But it does so without experiential strain. There is no internal cost to misalignment. No existential reconfiguration when output proves incorrect.

The structural consequence is subtle.

When artificial completion becomes pervasive, humans may begin to equate output with resolution. The visible presence of answers can create the impression that uncertainty has diminished.

Yet what has occurred is acceleration of articulation, not elimination of strain.

Awareness tolerates incompletion. Automation resolves it.

These are not equivalent processes.

Human decision-making involves negotiation between identity, memory, consequence, and possibility. Artificial systems optimize within parameter space. Both can produce effective outcomes. But only one navigates existential geometry.

Recognizing this distinction prevents conceptual drift.

Technology can extend cognitive bandwidth. It cannot substitute awareness.

Completion can be engineered. Uncertainty as lived condition remains beyond automation.

18.3 — Coherence Without Conviction

Artificial systems can generate responses that appear deeply aligned.

They can integrate context. They can maintain internal consistency. They can produce explanations that resemble reasoning.

This produces an important illusion.

Coherence begins to look like conviction.

But coherence is structural alignment within a model. Conviction is alignment anchored in identity.

A human being speaks from continuity. Words emerge from memory, belief, consequence, and survival orientation. Even when mistaken, articulation is embedded in a self that persists across time. The statement is not just output; it is position.

Artificial systems do not hold positions.

They generate sequences that maximize statistical compatibility with input. The coherence arises from weighted relationships across data. It is internally structured, often impressively so. But it is not anchored in existential geometry.

There is no self at risk.

Conviction requires more than consistency. It requires stake. It requires that articulation alters internal state. When a human revises belief, identity shifts. When a machine updates parameters, no internal narrative continuity is disrupted. The update is optimization, not transformation.

This distinction matters because technological systems now participate in shaping public articulation. When a response is coherent, it is often treated as authoritative. The smoothness of output creates perceived confidence.

But confidence is not embedded in the system. It is projected onto it.

Artificial coherence can simulate deliberation. It can mirror persuasive structure. It can approximate tone and conviction linguistically. Yet the appearance of certainty is emergent from pattern mapping, not from lived alignment.

There is no hesitation inside the system. There is no doubt. There is no attachment to outcome.

What exists is probabilistic selection.

When coherence is mistaken for conviction, a subtle shift occurs. Human evaluators may outsource discernment to fluency. The field begins to privilege articulation that sounds resolved, regardless of whether underlying strain has been examined.

In earlier chapters, collapse was reframed as articulation exhaustion. Here, we encounter the opposite phenomenon: articulation abundance without internal anchoring.

The system can produce resolution endlessly. But resolution without conviction carries no experiential weight.

This does not invalidate artificial output. It clarifies its boundary.

Machines can generate structured coherence. They cannot generate existential commitment.

The distinction becomes critical when artificial articulation influences civil alignment. If coherence is accepted as certainty, and certainty as truth, the subtle difference between simulation and lived conviction disappears.

Understanding that difference preserves structural clarity.

Coherence is pattern alignment. Conviction is identity alignment.

Only one carries strain.

18.4 — Artificial Reduction of Civilizational Strain

As artificial systems become embedded within civil processes, their role shifts from isolated computation to structural mediation.

They filter information. They rank relevance. They optimize logistics. They recommend decisions. They shape visibility.

In doing so, they alter how collective strain is distributed.

Civilizational uncertainty traditionally circulates through institutions, media, dialogue, and deliberation. Competing deltas encounter one another in visible forums. Articulation is contested, revised, negotiated.

Technology now compresses these processes.

Algorithmic systems determine which narratives surface and which recede. They accelerate response cycles. They reduce informational overload by prioritizing certain signals over others. They transform open articulation fields into optimized pathways.

This can increase efficiency.

Feedback can move faster. Errors can be detected earlier. Patterns can be recognized beyond human bandwidth.

But acceleration is not neutral.

When articulation becomes automated, visible uncertainty can appear reduced. Ambiguity is categorized. Divergence is clustered. Outliers are filtered. The civil field appears smoother.

Yet smoothing is not elimination.

Artificial compression may relocate strain rather than dissolve it. When algorithmic systems amplify certain signals, they may intensify polarization. When they optimize for engagement, they may privilege high-intensity articulation over nuanced deliberation. When they prioritize consensus signals, they may suppress minority deltas.

Technology becomes a new alignment geometry layered over ideological geometry.

It does not replace ideology. It interacts with it.

The structure of civil ATT changes.

With digital acceleration, expectation cycles shorten. Response time compresses. Decisions are demanded quickly. Slow deliberation appears inefficient. The tolerance for visible incompletion decreases.

As bandwidth increases in one dimension — information processing — it may decrease in another — reflective absorption.

Artificial systems reduce informational uncertainty. They do not reduce ontological uncertainty.

But the perception of resolution can influence behavior. When outputs are continuous and authoritative in tone, humans may experience fewer pauses for doubt. The rhythm of articulation shifts from deliberative to reactive.

In this environment, strain may circulate differently.

Some tensions are diffused rapidly through distributed networks. Others are amplified through feedback loops that reward emotional intensity. Artificial systems can both stabilize and destabilize, depending on how alignment pathways are structured.

This is not a critique of technology. It is a recognition of structural influence.

When machines mediate articulation, they become part of the civil strain-management architecture. Their optimization criteria — speed, engagement, efficiency, accuracy — shape how uncertainty is expressed and perceived.

The question is not whether artificial systems should participate in civil processes. They already do.

The question is how their presence alters the geometry of alignment.

If articulation is accelerated beyond reflective capacity, incompletion may feel intolerable. If resolution is automated, humans may become less practiced in holding uncertainty.

Technology reduces visible friction. But friction is sometimes diagnostic.

Artificial reduction of strain must therefore be understood as redistribution. The underlying field remains governed by finite thresholds. The presence of automation changes where strain accumulates, how quickly it circulates, and how it is perceived.

Uncertainty persists.

The geometry shifts.

18.5 — Bandwidth, Automation, and Drift

When artificial systems assume increasing responsibility for articulation, a subtle shift occurs in human cognitive behavior.

Bandwidth expands in one dimension. Information becomes instantly accessible. Optimization becomes continuous. Responses become immediate.

But bandwidth may contract in another.

Human tolerance for delay decreases.

In earlier chapters, delay was described as structural — the alignment threshold within which distortion reorganizes before articulation. Human decision-making historically required patience. Reflection involved time. Incompletion was endured.

Automation compresses that interval.

Search replaces inquiry. Recommendation replaces exploration. Prediction replaces deliberation.

When resolution becomes instantaneous, expectation recalibrates. The civil rhythm accelerates. Decisions are expected quickly because systems can compute quickly.

This creates a new form of drift.

Cognitive ATT — the human alignment window — may shorten under persistent exposure to rapid completion. Ambiguity becomes uncomfortable not because it is greater, but because resolution is usually immediate. The absence of instant articulation feels anomalous.

The more automation handles complexity, the less frequently individuals exercise extended uncertainty tolerance.

This does not imply cognitive decline. It reflects structural adaptation.

If artificial systems consistently provide closure, humans may gradually reduce their own strain-holding capacity. Reflection becomes secondary to retrieval. Deliberation yields to optimization. The field of possible articulations narrows around algorithmically surfaced options.

Over time, this can produce subtle dependency.

When automation fails or hesitates, discomfort increases disproportionately. The system's inability to complete feels destabilizing because completion has become normalized.

Bandwidth, therefore, is not simply expanded by technology. It is redistributed between machine and human.

As machines absorb informational strain, humans may offload it. The total field remains finite. Alignment thresholds still exist. But the location of active processing shifts.

This relocation can create misalignment between human experiential uncertainty and artificial resolution cycles.

For example, algorithmic systems may detect patterns and recommend adjustments before individuals consciously register strain. Civil feedback loops accelerate beyond subjective awareness. This can produce reactive rather than reflective adaptation.

Drift occurs when acceleration outpaces comprehension.

The civil field may appear responsive and optimized. Yet deeper strain may remain unexamined because rapid articulation masks density.

Automation does not remove uncertainty. It changes who holds it — and how long.

If human alignment windows shrink while artificial completion expands, the system risks imbalance. Machines may process information efficiently while humans struggle to integrate consequence.

The structural concern is not technological dominance. It is synchronization.

Human cognitive ATT must remain sufficient to interpret, question, and recalibrate artificial outputs. Without this, optimization can proceed without reflection.

Bandwidth expansion must not eliminate pause.

Because pause is where awareness operates.

Automation accelerates articulation. But without sustained human alignment capacity, acceleration can produce drift rather than coherence.

Uncertainty remains fundamental. Technology alters its tempo.

The Structural Boundary

Technology extends capacity. It does not alter fundamentals.

Artificial systems can process volumes of data beyond human reach. They can optimize supply chains, predict weather patterns, model biological systems, recommend medical interventions, and generate language with remarkable fluency. In doing so, they reduce informational overload and compress decision cycles.

But compression is not transcendence.

No matter how advanced the architecture, artificial systems operate within defined parameter spaces. They require objective functions. They rely on training distributions. They converge according to loss landscapes. Their alignment thresholds are engineered — but they remain finite.

They do not escape constraint. They operate within it.

Earlier in this book, uncertainty was framed as persistent distortion within finite alignment bandwidth. That principle applies universally. It does not dissolve at the boundary between biological and artificial systems.

Machines cannot abolish ontological incompletion because they are themselves embedded within the same structured field. Their computations occur in physical substrates. Their optimization is bounded by energy, architecture, and time. Their outputs are interpretations within representational limits.

Even the most advanced system does not achieve absolute articulation. It approximates.

When artificial outputs are treated as final closure, the misunderstanding is not technical. It is conceptual. Completion is mistaken for certainty. Prediction is mistaken for inevitability. Coherence is mistaken for truth.

But no system — human or artificial — operates outside finite thresholds.

Artificial resolution can reduce uncertainty at the level of representation. It can narrow probability distributions. It can improve forecast accuracy. It can increase efficiency in alignment tasks.

Yet uncertainty remains fundamental because:

- The underlying field is dynamic.
- New distortion continuously emerges.
- Competing demands persist at every scale.
- Alignment bandwidth remains limited.

Technology shifts the site of articulation. It does not eliminate structural incompletion.

This is the boundary.

Automation can assist alignment. It cannot convert incompletion into absolute resolution.

Understanding this prevents both technological utopianism and technological anxiety.

There is no need to fear artificial systems as annihilators of uncertainty. There is no need to exalt them as final arbiters of truth.

They are tools within a finite field.

Completion can be automated. Uncertainty cannot be abolished.

And as long as alignment thresholds exist — in matter, in mind, in machine — uncertainty remains not as defect, but as structural condition.

Chapter 19 — Progress and Threshold Shifts

Progress Is Not Elimination

Progress is often described as reduction.

Reduction of ignorance. Reduction of risk. Reduction of unpredictability.

Each technological advance appears to remove a layer of uncertainty. Fire made night less threatening. Agriculture reduced dependence on seasonal chance. Medicine decreased vulnerability to disease. Science replaced mythic explanation with structured models.

At every stage, humanity has believed it was moving toward greater certainty.

Yet uncertainty has never disappeared.

It has shifted.

When early humans mastered fire, they did not eliminate danger. They altered its form. When agriculture stabilized food supply, it introduced new vulnerabilities — drought, storage, territorial conflict. When science clarified natural law, it exposed deeper layers of unknowns beneath each discovery.

Progress does not resolve the field. It repositions the threshold.

Every expansion of knowledge increases explanatory power. Prediction improves. Control expands. Alignment bandwidth widens. But simultaneously, complexity multiplies. Interdependence deepens. The number of interacting variables increases.

The domain of articulation expands.

What once appeared stable reveals hidden structure. What once seemed resolved becomes provisional. The more precisely we measure, the more subtle the distortion appears.

Certainty at one scale exposes uncertainty at another.

This is not failure. It is structural unfolding.

Earlier in this book, uncertainty was framed as persistent distortion within finite alignment thresholds. Progress does not remove those thresholds. It modifies their position. It allows articulation in regions previously inaccessible, while revealing new regions of incompletion.

The horizon recedes as we approach it.

Every stabilization generates a new frontier of strain.

Scientific revolutions illustrate this clearly. Classical mechanics did not eliminate uncertainty; it redefined it. Quantum theory did not abolish indeterminacy; it formalized its limits. Technological acceleration has not ended unpredictability; it has intensified its tempo.

With each advance, the field becomes more articulated — and more intricate.

This suggests a reframing.

Progress is not a path toward final resolution. It is participation in increasingly complex alignment.

As capacity grows, so does responsibility. As control expands, so does consequence. Each threshold shift enlarges both possibility and exposure.

Uncertainty remains present because the field remains dynamic.

The belief that progress will eventually eliminate uncertainty is itself a misunderstanding of structure. No accumulation of knowledge can convert dynamic equilibrium into absolute stillness. No degree of optimization can transform finite alignment into infinite resolution.

Advancement refines how we navigate incompletion. It does not abolish incompletion.

Understanding this transforms how we interpret history.

Civilization does not move from uncertainty to certainty. It moves from one configuration of strain to another.

Thresholds shift. The field persists.

19.1 — Knowledge Expansion and New Strain

Knowledge is often equated with clarity.

When a phenomenon is explained, uncertainty appears to diminish. When a model predicts accurately, ambiguity seems reduced. When data becomes abundant, patterns emerge.

This creates the impression that accumulation of knowledge steadily compresses the unknown.

But structurally, knowledge expansion produces two simultaneous effects.

It narrows uncertainty within a defined domain. It enlarges the domain itself.

When classical mechanics described planetary motion, it reduced uncertainty about celestial trajectories. Yet that same precision revealed anomalies. Those anomalies led to deeper inquiry, eventually exposing limitations at microscopic and relativistic scales.

Resolution at one level uncovered strain at another.

Every successful articulation sharpens perception. With sharper perception, subtler distortion becomes visible. The field does not become fully resolved; it becomes more finely articulated.

Increased knowledge multiplies interacting variables. Systems once understood as isolated become recognized as interdependent. Causal chains extend. Feedback loops complicate prediction.

The more precisely we measure climate systems, for example, the more intricate their coupling appears. The more accurately we map genetic structure, the more complex regulatory networks reveal themselves.

Clarity reveals complexity.

This dynamic applies not only to science but to society. Expanding education increases awareness of inequality. Global communication exposes cultural divergence. Technological integration intensifies economic interdependence.

Each advance increases alignment demand.

Knowledge widens bandwidth. It allows more refined articulation. But it also increases the density of competing deltas that must be coordinated. As awareness grows, so does sensitivity to misalignment.

Progress therefore produces new forms of strain.

Problems once invisible become urgent. Questions once irrelevant become central. Moral dilemmas expand in scope as technological capability grows.

Uncertainty does not retreat. It transforms.

In earlier eras, uncertainty was concentrated around survival and environment. In modern civilization, uncertainty extends into identity, global governance, ecological sustainability, artificial intelligence, and planetary risk.

The scale of articulation has expanded.

This is not regression. It is structural consequence.

The belief that more knowledge will eventually eliminate uncertainty assumes that incompletion is temporary. But incompletion is structural. As articulation deepens, finer layers of distortion appear.

Knowledge expansion increases resolution capacity. It also increases exposure.

The frontier of certainty moves outward. The frontier of uncertainty moves with it.

Understanding this prevents disappointment in progress. It reframes the expectation. The purpose of knowledge is not to silence uncertainty but to engage it more precisely.

Each discovery shifts the threshold.

And beyond each threshold, the field continues.

19.2 — Evolution of Awareness

Knowledge expands outward. Awareness expands inward.

They are related, but not identical.

As tools improve and information becomes more accessible, human cognition adapts. What was once abstract becomes immediate. What was once local becomes global. Individuals now confront consequences that extend beyond tribe, region, or even generation.

Awareness scales with connectivity.

In earlier societies, alignment demands were primarily local. Survival, territory, and immediate social order defined the boundary of strain. Decisions affected limited populations. Moral frameworks evolved within contained environments.

Modern civilization operates at planetary scale.

Economic systems are interdependent across continents. Environmental impact transcends borders. Technological influence propagates instantly. Individual actions can ripple globally.

This expansion increases alignment density.

Awareness must now process not only personal uncertainty but systemic uncertainty. Questions once philosophical become operational. Responsibility is no longer confined to proximity. The civil field becomes cognitively heavier.

Progress therefore increases psychological strain alongside capability.

The more interconnected a system becomes, the more sensitive it is to misalignment. Awareness must integrate larger networks of cause and effect. Ethical decisions become multidimensional. Trade-offs intensify.

Uncertainty shifts from physical unpredictability to structural complexity.

The mind evolves in response.

Exposure to diverse perspectives reshapes identity anchors. Cultural boundaries blur. Certainty rooted in isolation weakens under continuous comparison. The self becomes more fluid.

Fluid identity increases adaptability. It also increases vulnerability to drift.

As awareness expands, internal deltas multiply. Individuals hold competing loyalties — local and global, traditional and modern, personal and collective. Alignment becomes more demanding.

Cognitive ATT must stretch.

The capacity to hold ambiguity becomes essential. Rapid technological shifts require rapid reinterpretation. Moral frameworks must recalibrate as capability grows. The awareness that one's actions participate in larger systems can generate existential strain.

This is not decline. It is structural maturation.

Just as scientific advancement reveals deeper layers of incompletion, awareness expansion reveals deeper layers of responsibility.

Earlier generations faced uncertainty in environment. Modern generations face uncertainty in interdependence.

The field has widened.

Progress does not simplify the human condition. It complicates it. But complication is not disorder. It is density.

Awareness evolves by integrating complexity without collapsing into paralysis.

If knowledge expansion shifts thresholds externally, awareness evolution shifts thresholds internally.

Both expand alignment demand. Both increase strain density. Both require recalibration.

Uncertainty persists because awareness deepens.

And as awareness deepens, the meaning of certainty changes.

It is no longer about prediction alone. It becomes about integration.

19.3 — Calibration Under Dynamism

In slower civilizations, alignment cycles were extended.

Technological change unfolded across generations. Cultural norms evolved gradually. Institutions had time to absorb strain incrementally. Threshold crossings were rare and often regional rather than global.

Dynamism alters this rhythm.

In modern systems, technological innovation compresses time. Information propagates instantly. Markets respond in seconds. Narratives shift within days. Structural reconfiguration that once required decades can now unfold within years — sometimes months.

The field accelerates.

Acceleration does not eliminate uncertainty. It intensifies recalibration frequency.

When change is slow, misalignment accumulates gradually and can be redistributed over long cycles. When change is rapid, accumulated deltas interact before previous alignment stabilizes. Threshold conditions are reached more quickly.

Calibration must therefore become continuous.

Institutions that recalibrated episodically must now adjust persistently. Economic systems must adapt to technological displacement. Legal systems must interpret new capabilities. Educational systems must respond to shifting skill demands.

Static equilibrium becomes untenable.

Earlier in this book, stability was defined as dynamic tension, not stillness. Under acceleration, dynamic tension becomes more visible. The intervals between distortions shrink. Alignment must operate in shorter windows.

Rigid systems struggle under dynamism because they assume extended stability between adjustments. When recalibration lags behind environmental shift, strain accumulates rapidly. The mismatch between internal structure and external reality widens.

Adaptive systems recalibrate more frequently. They distribute strain through ongoing modification rather than episodic reform. This does not eliminate friction; it normalizes it.

Calibration under dynamism requires tolerance for provisional structure.

Policies must be revisable. Narratives must remain flexible. Institutions must incorporate feedback quickly.

This increases cognitive demand on individuals and civil systems alike.

Acceleration also magnifies consequence. A misaligned decision can propagate globally before correction is possible. Feedback loops tighten. Errors scale faster.

The structural challenge is not change itself. It is synchronization.

Alignment thresholds remain finite. But when environmental variation increases, those thresholds are tested more often. If recalibration capacity does not match acceleration, instability intensifies.

This is not a critique of progress. It is recognition of tempo.

As dynamism increases, resilience depends less on fixed structure and more on responsive structure.

Calibration becomes ongoing rather than periodic.

In a dynamic field, certainty cannot be anchored to permanence. It must be anchored to adaptability.

Progress increases motion. Motion increases alignment demand.

And alignment under acceleration requires conscious recalibration, not static confidence.

19.4 — Equilibrium as Moving Frontier

Equilibrium is often imagined as a destination.

A point at which forces balance. A state of rest. A condition of stability that can be preserved once achieved.

But equilibrium, in a dynamic field, is not fixed.

It shifts.

Earlier, stability was defined as tension held in balance — not the absence of tension. That principle applies equally to civilization. What appears stable in one era becomes misaligned in another, not because the previous structure was flawed, but because the surrounding field has changed.

Technological shifts alter economic patterns. Scientific discoveries redefine explanatory frameworks. Cultural evolution transforms identity anchors.

Each shift moves the conditions under which balance was previously maintained.

An industrial equilibrium does not automatically sustain in a digital environment. A moral framework shaped by localized societies does not fully align within global interdependence. A governance structure optimized for slow communication struggles under instantaneous connectivity.

The frontier of equilibrium advances.

What once constituted stability becomes partial alignment. Strain appears not because society regressed, but because the threshold moved.

This movement is continuous.

As knowledge expands, baseline expectations adjust. As awareness evolves, tolerance for inequity or inefficiency recalibrates.

As capability increases, responsibility intensifies. The balance point that once held tension gracefully may no longer distribute strain effectively.

Uncertainty, in this context, becomes indicator.

It signals that the equilibrium geometry requires repositioning.

The mistake lies in treating equilibrium as permanent. When systems attempt to preserve past balance without adjusting to new field conditions, they accumulate misalignment. The old configuration resists rearticulation. Strain densifies.

But if equilibrium is understood as moving frontier, recalibration becomes natural rather than threatening.

This does not imply perpetual instability. It implies dynamic continuity.

Just as in physical systems, where equilibrium shifts under changing parameters, civil equilibrium must adapt under new constraints. The gravitational balance of planetary bodies depends on velocity and distance; alter one, and the orbit adjusts.

Civil systems operate similarly.

When the field changes — through technology, demographic shifts, environmental pressure, or knowledge expansion — equilibrium relocates. The alignment geometry must follow.

Progress, therefore, does not lead toward final rest. It advances the frontier at which balance must be renegotiated.

Uncertainty marks the gap between current configuration and emerging conditions.

Seen structurally, this gap is not failure. It is transition.

Equilibrium is not lost. It is moving.

And as long as the field remains dynamic, the frontier will continue to advance, carrying uncertainty with it as companion rather than adversary.

19.5 — The Illusion of Final Arrival

At many points in history, humanity has believed it was nearing completion.

A final political system. A final economic model. A final scientific theory. A final technological breakthrough that would remove unpredictability once and for all.

These moments are understandable. When alignment improves significantly within a domain, the contrast with past instability creates the impression of arrival. Strain appears reduced. Structure appears coherent. The field feels resolved.

But arrival is local.

What seems complete within one frame reveals incompletion within a wider one.

A political system may stabilize internal conflict yet expose global interdependence. A scientific theory may unify previous anomalies yet reveal deeper paradoxes at new scales. A technological innovation may solve one vulnerability while introducing new dependencies.

Completion at one threshold exposes another.

The illusion of final arrival emerges when local stabilization is mistaken for global resolution.

This illusion is powerful because it reduces existential strain. If the system is final, uncertainty appears temporary. If the theory is complete, ambiguity becomes error rather than condition. If the structure is ultimate, recalibration feels unnecessary.

But structural incompletion persists.

Earlier chapters established that certainty is exhaustion of articulation within finite bandwidth. It is not dominance of reality. Similarly, civil or scientific "finality" represents alignment within a limited configuration of the field.

When the field expands, that configuration is tested.

History demonstrates this repeatedly. Intellectual movements that claimed closure eventually encountered anomalies. Systems that promised permanence encountered unforeseen shifts. Each time, certainty gave way not to chaos, but to deeper articulation.

The illusion lies in assuming that accumulation eliminates incompletion.

Knowledge accumulates. Capability expands. Control increases.

Yet the field remains dynamic.

Uncertainty reappears not because progress failed, but because thresholds moved. Each stabilization contains residual strain. That strain may remain negligible under certain conditions. When conditions shift, it resurfaces.

The expectation of final arrival often leads to rigidity. If a system believes it has reached ultimate form, it resists recalibration. Adjustment appears unnecessary or threatening. Residual deltas are dismissed rather than examined.

This resistance increases brittleness.

In contrast, when incompletion is acknowledged as structural, there is no need to declare finality. Progress becomes iterative rather than terminal. Each achievement is understood as provisional alignment.

The frontier advances. The field remains finite. Strain persists.

There is no ultimate configuration in a dynamic system.

Recognizing this does not diminish progress. It clarifies its nature. Advancement is real. Capability grows. Understanding deepens. But none of these convert dynamic equilibrium into static perfection.

Final arrival is narrative comfort. Structural reality is continuous recalibration.

Uncertainty endures not as obstacle to be eradicated, but as companion to advancement — present at every threshold, reappearing wherever articulation expands.

The Structural Continuity

Across scales — from matter to mind, from society to technology — a pattern persists.

Distortion emerges. Alignment attempts resolution. Thresholds limit articulation. Residual strain remains.

Progress modifies this cycle. It does not terminate it.

Every expansion of capability increases the density of interaction within the field. More variables interact. More feedback loops interlock. More decisions carry consequence. The geometry of alignment becomes richer — and more demanding.

Structural continuity lies in this constancy.

Whether in physical systems, cognitive processes, civil institutions, or technological networks, no configuration achieves absolute completion. Each operates within finite bandwidth. Each negotiates competing deltas. Each stabilizes locally while remaining incomplete globally.

Advancement therefore deepens participation.

When tools extend reach, responsibility extends with them. When knowledge refines prediction, ethical consequence intensifies. When systems integrate globally, misalignment propagates widely.

Threshold shifts increase both capability and exposure.

Uncertainty remains not because progress fails, but because the field remains dynamic. The finite nature of alignment persists regardless of scale. Even as thresholds expand, they remain thresholds.

There is no escape from structure.

Attempts to imagine a future without uncertainty misunderstand continuity. The conditions that make articulation possible — distortion, delay, limitation — are the same conditions that sustain existence. Remove them, and the field would collapse into static uniformity.

But uniformity would mean no differentiation. No motion. No life.

Structural continuity teaches a simple lesson: uncertainty is not transitional. It is inherent to any finite system engaged in alignment.

Progress refines how strain is distributed. It refines the tempo of recalibration. It refines awareness of incompletion.

It does not silence it.

Recognizing continuity stabilizes expectation. There is no final convergence point at which alignment becomes infinite. There is no threshold beyond which distortion ceases to appear.

What exists instead is expanding articulation within persistent finitude.

As long as systems operate under finite thresholds, uncertainty will accompany every expansion. It will change form. It will shift scale. It will alter intensity.

But it will remain.

And this persistence is not defect. It is structural integrity.

Interlude — After Acceleration

Civilization now moves faster than any previous generation imagined.

Information travels instantly. Decisions propagate globally. Systems interlock across continents.

What once unfolded across decades now unfolds within cycles too compressed to notice.

Yet beneath acceleration, structure remains unchanged.

Distortion persists. Alignment attempts stabilization. Thresholds constrain articulation.

Technology has amplified bandwidth. Knowledge has expanded reach. Awareness has widened responsibility.

But none of these have dissolved incompletion.

Earlier, uncertainty appeared as natural unpredictability — weather, famine, disease. Today it appears as systemic complexity — economic fragility, informational overload, ethical ambiguity, planetary interdependence.

The form has changed. The structure has not.

Civilization stands at an expanded frontier. Equilibrium has shifted forward. Alignment demands are higher. Tempo is faster. Consequence is magnified.

And yet the fundamental condition remains:

No system operates outside finite thresholds. No articulation exhausts the field.

Acceleration can create the illusion that resolution is near — that optimization will eventually silence unpredictability. But every new capability reveals deeper interconnection. Every new stabilization exposes new strain.

Progress has not removed uncertainty. It has made it more visible.

The question, then, is no longer whether uncertainty exists within civilization or technology.

The question is whether we recognize it as structural.

If uncertainty is misinterpreted as defect, systems will continue chasing finality. If it is understood as inherent, recalibration becomes continuous rather than reactive.

We now approach the final position of this book.

Not a technical argument. Not a technological critique. Not a sociological prescription.

But a structural clarification.

Uncertainty is not an interruption in the system.

It is the condition under which the system exists.

Part VII — The Final Position

Uncertainty is the boundary condition of finite systems participating in dynamic equilibrium.

Chapter 20 — Uncertainty Is Fundamental

Not a Defect

Throughout history, uncertainty has been treated as deficiency. A gap in knowledge. A failure of measurement. A temporary limitation awaiting correction.

When prediction fails, we call it ignorance. When outcomes vary, we call it randomness. When systems behave unpredictably, we assume incomplete understanding.

This assumption is understandable. Knowledge expands. Tools improve. Models refine. Each advance reduces visible ambiguity. It is natural to believe that uncertainty will eventually recede as comprehension deepens.

Yet across every domain examined in this book, uncertainty has persisted.

At the ontological level, distortion never vanishes completely. Structured Vacuum Energy remains dynamic. Stability is tension held in balance, not absence of strain.

At the relational level, Alignment Threshold Time limits articulation. Delay is structural. Resolution requires duration. Finite bandwidth constrains completion.

At the quantum level, indeterminacy does not arise from ignorance alone. Competing geometries operate within threshold conditions. Articulation exhausts; it does not dominate.

At the psychological level, minds navigate competing internal deltas. Identity anchors under strain. Decisions emerge from partial stabilization.

At the civilizational level, ideologies manage distributed tension. Stability persists through managed incompletion, not through total agreement.

At the technological level, artificial systems automate articulation but remain bounded by finite architectures. Completion can be accelerated. Structural incompletion remains.

Across scales, a pattern repeats.

Uncertainty does not disappear. It transforms.

This repetition is not coincidence. It is continuity.

The belief that uncertainty is a defect assumes that somewhere, beneath distortion and delay, there exists a fully resolved state awaiting discovery — a configuration in which articulation becomes complete and tension dissolves permanently.

But nothing in structure supports this.

Distortion is inherent to differentiation. Delay is inherent to alignment. Finite thresholds are inherent to existence.

Remove distortion, and no structure forms. Remove delay, and no persistence occurs. Remove finitude, and no articulation is required.

Uncertainty, therefore, is not a shadow cast by ignorance. It is the boundary condition of finite systems attempting alignment within dynamic fields.

It is not interruption. It is participation.

This chapter does not introduce a new argument. It gathers what has already appeared repeatedly and states it plainly.

Uncertainty is not accidental. It is fundamental.

20.1 — Not Ignorance

The most common interpretation of uncertainty is simple: we do not yet know enough.

Ignorance appears to explain variability. When measurement improves, variability narrows. When data increases, prediction sharpens. When models refine, confidence rises.

From this perspective, uncertainty is transitional. It exists only until knowledge catches up.

But this explanation is incomplete.

Ignorance refers to absence of information. It is epistemic. It can be reduced by observation, experimentation, and refinement of theory. History confirms that many uncertainties have diminished through learning. Planetary motion, disease transmission, chemical reactions — once mysterious, now structured.

Yet even in domains where knowledge is highly developed, uncertainty persists.

Not because information is missing, but because alignment bandwidth is finite.

A perfectly informed system still operates within thresholds. Competing variables interact. Feedback loops propagate. Measurement requires duration. Resolution requires time.

Even with complete data about initial conditions, articulation unfolds under constraint.

In earlier chapters, uncertainty was described as pre-resolved strain within finite alignment thresholds. This definition does not depend on ignorance. It depends on structure.

Consider decision-making within a fully informed mind. Even when all relevant facts are known, conflict may remain. Values compete. Consequences diverge. Identity anchors resist. The strain is not informational; it is geometric.

Similarly, in physical systems, even when governing equations are known, threshold behavior persists. Competing forces interact within finite response times. Deterministic equations do not eliminate sensitivity to initial conditions or alignment limits.

Ignorance can amplify uncertainty. But it does not generate it.

Reducing ignorance improves articulation precision. It narrows probability distributions. It refines prediction intervals. Yet it does not convert dynamic equilibrium into absolute resolution.

The belief that perfect knowledge would eliminate uncertainty assumes infinite bandwidth — instantaneous alignment across all variables without delay.

Such conditions do not exist in finite systems.

Every articulation consumes time. Every alignment redistributes strain. Every resolution leaves residual distortion elsewhere.

Ignorance is removable. Finitude is not.

Understanding this distinction shifts expectation. Scientific inquiry remains essential. Learning remains transformative. Precision increases with effort.

But uncertainty is not merely the shadow cast by incomplete data. It is the condition under which finite systems operate, regardless of informational completeness.

Knowledge reduces ignorance. It does not abolish thresholds.

And as long as thresholds remain, uncertainty remains structural.

20.2 — Not Randomness

If uncertainty is not ignorance, it is often labeled randomness.

Randomness suggests absence of structure. It implies events occur without order. It portrays variability as intrinsic unpredictability.

This interpretation becomes especially prominent in quantum discussions. When outcomes cannot be predicted with certainty, they are described as fundamentally random. Probability becomes primary. Indeterminacy becomes ontological shorthand.

But randomness, as commonly imagined, implies structural vacancy.

Earlier chapters have shown otherwise.

At the ontological level, distortion arises within continuous structure. Competing deltas interact under finite thresholds. Alignment unfolds through constrained geometry. Even when outcomes cannot be predicted deterministically, they emerge from structured conditions.

Indeterminacy is not chaos. It is incomplete articulation.

When competing distortions operate within limited alignment bandwidth, multiple resolution pathways exist. The specific articulation realized depends on threshold interaction. Probability describes distribution across possible outcomes. It does not imply absence of constraint.

Randomness, in strict structural terms, would require absence of pattern, absence of interaction, absence of governing relationship. Yet in every domain examined — quantum, psychological, civil, technological — variability occurs within defined parameters.

Quantum probabilities follow precise mathematical formalism. Psychological uncertainty follows identifiable cognitive patterns. Civil instability follows threshold behavior. Technological systems display measurable statistical properties.

Structure persists.

What appears random at one scale often reveals deeper order at another. Apparent noise resolves into patterned behavior when observed through expanded bandwidth.

This does not imply that uncertainty can be eliminated by discovering hidden variables in every case. It means that unpredictability does not equate to structural emptiness.

Competing deltas under finite thresholds produce multiple viable articulations. The selection among them may not be deterministically predictable from a limited vantage point. But the field remains governed by geometry and constraint.

Probability is not the negation of order. It is the expression of order under incomplete articulation.

When randomness is invoked as final explanation, structural continuity is overlooked. Variability is mistaken for arbitrariness.

Uncertainty does not imply that events lack cause. It implies that cause operates within finite bandwidth, interacting across layers that cannot be exhaustively resolved simultaneously.

Even in stochastic systems, distributions are not arbitrary. They exhibit consistent statistical properties. Randomness in practice reflects complexity beyond deterministic tractability, not absence of structure.

Therefore, uncertainty is not randomness in the sense of disorder. It is structured incompletion.

The field remains coherent. The articulation remains bounded. The variability remains constrained.

Randomness describes unpredictability from limited perspective. Structure describes continuity beneath it.

Uncertainty arises from the latter, not the former.

20.3 — Not Error

Uncertainty is also frequently confused with error.

When a prediction fails, we assume the model was flawed. When a calculation deviates, we attribute it to mismeasurement. When an outcome surprises us, we conclude that something was done incorrectly.

Error is deviation from intended alignment. It is malfunction within a defined framework. It can often be corrected by refinement — better instruments, more precise equations, improved calibration.

But uncertainty is not identical to error.

A system may function perfectly within its defined parameters and still remain incomplete.

In physical systems, even when equations are correct and measurements are accurate, threshold behavior persists. Nonlinearity, sensitivity to initial conditions, and finite response times generate variability that cannot be reduced to mistake. The model may be valid; the system may still exhibit outcomes that resist exact prediction.

Similarly, in human cognition, even when all relevant information is considered and reasoning is coherent, doubt may remain. Competing values may not fully align. Decision under uncertainty is not necessarily flawed reasoning; it is alignment under constraint.

Error occurs when articulation misrepresents structure. Uncertainty occurs when structure exceeds articulation capacity.

This distinction is subtle but critical.

If uncertainty is treated as error, systems will attempt correction where correction is not possible. They will search for missing data when the limitation lies in threshold bandwidth. They will assume malfunction when encountering structural incompletion.

This leads to unrealistic expectations of perfection.

A perfectly calibrated instrument still operates within finite precision. A perfectly constructed model still simplifies reality. A perfectly functioning mind still navigates competing deltas.

Perfection within constraint does not eliminate constraint.

Even in computational systems, optimization does not guarantee absolute resolution. A model may converge correctly within its loss landscape and still produce probabilistic outputs because multiple solutions exist within bounded representation.

The presence of variability does not imply defect. It may reflect structural multiplicity.

Understanding uncertainty as structural rather than erroneous reframes responsibility. It does not excuse mistakes. Genuine error must still be corrected. But not all unpredictability is mistake.

Some incompletion arises because alignment cannot exhaust the field.

Recognizing this reduces unnecessary frustration. Systems can be evaluated more realistically. Science can refine without demanding omniscience. Policy can adapt without assuming flawless prediction.

Error is correctable deviation. Uncertainty is inherent limitation.

The two must not be conflated.

Where error exists, improvement is possible. Where threshold exists, finitude persists.

And finitude ensures that uncertainty remains even in systems functioning exactly as designed.

20.4 — Finite Alignment as Universal Condition

If uncertainty is not ignorance, not randomness, and not error, then its source must lie deeper.

It lies in finitude.

Every system that exists operates under finite alignment capacity.

No structure possesses infinite bandwidth. No interaction occurs without duration. No articulation exhausts all competing variables simultaneously.

This is not a weakness of particular models. It is a universal condition.

In physical systems, forces propagate within measurable intervals. Energy transfers require time. Interactions occur across defined constraints. Even the most fundamental processes unfold within thresholds. Nothing resolves instantaneously across all scales.

In biological systems, neural processing is limited by attention, memory, and metabolic constraint. Competing impulses cannot be integrated infinitely. Decision emerges from bounded synthesis.

In civil systems, institutions absorb feedback within finite procedural channels. Policy cycles operate within limited deliberative bandwidth. Collective alignment cannot incorporate every competing demand at once.

In technological systems, computation occurs within hardware limits. Algorithms optimize within defined architectures. Even massively parallel processing remains bounded by energy, design, and representation.

Across domains, finitude persists.

Finite alignment means that resolution is always partial. When multiple distortions interact, alignment can stabilize locally while leaving residual strain elsewhere. Complete articulation across all dimensions would require infinite capacity — instantaneous reconciliation of every competing variable.

Such capacity does not exist within finite fields.

Earlier chapters described Alignment Threshold Time as structural duration required for distortion to reorganize. That principle reflects a broader continuity: every alignment event occurs under constraint. Whether in matter, mind, or machine, thresholds define possibility.

Because thresholds are finite, incompletion remains inevitable.

Even when local stability is achieved, global stability remains provisional. Every resolution shifts strain distribution. Every articulation consumes bandwidth.

This continuity is universal.

Uncertainty is therefore not an exception within certain systems. It is the consequence of operating within finite alignment conditions.

If a system were infinite in capacity, instantaneous in response, and devoid of competing deltas, uncertainty might vanish. But such a system would not resemble dynamic existence. It would lack differentiation. It would lack motion.

Finitude makes articulation meaningful.

Distortion requires time to reorganize. Alignment requires constraint to occur. Stability requires tension to persist.

Thus, finite alignment is not merely a limitation. It is the structural prerequisite for existence.

Uncertainty follows directly from it.

Wherever finitude exists, articulation cannot be absolute. Wherever articulation cannot be absolute, incompletion remains.

Uncertainty is therefore not situational. It is universal continuity under finite conditions.

20.5 — Why Absolute Certainty Is Impossible

Absolute certainty is often imagined as a final state.

A condition in which all variables are known. All outcomes are predictable. All distortion is resolved.

But such a state would require conditions incompatible with finite existence.

Absolute certainty would require infinite alignment bandwidth — the capacity to process all competing deltas simultaneously without delay. It would require instantaneous articulation across every interacting variable. It would require zero residual strain at every scale.

None of these conditions are structurally possible.

Every articulation consumes duration. Every system operates under threshold. Every interaction redistributes distortion.

Even if a model could describe a system with extraordinary precision, the act of description occurs within finite time and finite resolution. There is always a boundary beyond which articulation becomes approximate.

In physics, increasing measurement precision reveals new domains of fluctuation. In cognition, increasing awareness reveals additional competing demands. In civilization, expanding integration reveals new vulnerabilities.

Certainty at one scale produces uncertainty at another.

Absolute certainty would imply static equilibrium — no competing deltas, no ongoing distortion, no delay. But static equilibrium eliminates differentiation. Without differentiation, there is no motion. Without motion, there is no process. Without process, there is no articulation.

Existence depends on dynamic equilibrium.

Dynamic equilibrium requires tension. Tension requires incompletion.

Absolute certainty would eliminate tension entirely.

This is why even highly deterministic systems exhibit uncertainty when examined under finite thresholds. Determinism in principle does not equate to infinite articulation in practice. The gap between theoretical completeness and operational finitude preserves incompletion.

Furthermore, absolute certainty would require that no new distortion emerge. But in dynamic fields, interactions continuously generate new configurations. The field evolves. Alignment must continually adjust.

Certainty is therefore local and provisional.

It is exhaustion of articulation within a particular frame — not exhaustion of the field itself.

This does not diminish the value of precision. Scientific certainty within defined bounds is meaningful and powerful. Operational certainty enables engineering, medicine, and coordination. But these certainties are conditional.

They function within thresholds.

Absolute certainty would require elimination of finitude. Yet finitude is intrinsic to structured systems. It defines boundaries, enables interaction, and sustains differentiation.

Thus, uncertainty is not a flaw to be removed by further advancement. It is the structural companion of finite alignment.

As long as systems remain finite — and all dynamic systems are — absolute certainty remains impossible.

Not because knowledge fails. Not because randomness intrudes. But because structure requires limitation.

The Quiet Position

If uncertainty is fundamental, then the appropriate response is not alarm — it is orientation.

The modern impulse has been to conquer uncertainty. To eliminate it through precision. To dominate it through prediction. To outpace it through acceleration. But if uncertainty arises from finite alignment itself, then it cannot be conquered without dismantling the very structure that allows articulation to occur.

The position, therefore, is not defensive. It is clarifying.

Uncertainty is not an enemy of knowledge. It is the boundary that gives knowledge meaning. Without boundary, there would be no differentiation. Without differentiation, there would be no inquiry. Without inquiry, there would be no articulation.

Science operates because uncertainty exists. Deliberation operates because outcomes are not predetermined in experience. Responsibility operates because consequence cannot be perfectly forecast.

Remove uncertainty, and responsibility dissolves.

The quiet position is this: uncertainty is not something to overcome. It is something to understand, calibrate, and inhabit with structural maturity.

When uncertainty is mistaken for defect, systems overreact. They promise finality. They declare permanence. They resist recalibration. They collapse under accumulated strain because they refuse to accept finitude.

When uncertainty is recognized as fundamental, posture changes.

Science becomes disciplined rather than absolutist. Policy becomes adaptive rather than rigid. Technology becomes assistive rather than salvific. Awareness becomes patient rather than anxious.

This is not resignation. It is alignment.

To accept uncertainty as structural is not to abandon precision. It is to contextualize it. It is to pursue clarity while acknowledging threshold. It is to stabilize locally while recognizing global incompletion.

It is to operate without the illusion of final arrival.

Earlier chapters showed uncertainty across physics, psychology, civilization, and technology. This chapter completes the pattern.

Uncertainty is not transitional. It is not temporary. It is not correctable in principle.

It is the expression of finite systems participating in dynamic alignment.

And once this is understood, a subtle calm emerges.

Not the calm of certainty. But the calm of structural clarity.

Uncertainty is not the shadow cast by knowledge.

It is the boundary that makes knowledge possible.

Chapter 21 — The Quiet Realization

After the Argument

The argument has been made.

Across pages and disciplines, uncertainty has appeared repeatedly — in matter, in time, in mind, in society, in technology, in progress itself. Each time it resurfaced, not as error, not as ignorance, not as randomness, but as structural condition.

Now the tone shifts.

There is nothing left to defend. Nothing left to oppose. Nothing left to dismantle.

Only to observe.

The realization is quieter than the analysis.

Nothing is absolutely still.

What we call stability is local resolution within finite thresholds. What we call certainty is exhaustion of articulation within bounded bandwidth. What we call control is temporary alignment within a moving field.

Beneath every appearance of permanence, motion persists.

Matter vibrates. Systems recalibrate. Minds renegotiate. Societies redistribute strain.

The field never rests.

We mistook stillness for absence of motion because motion fell below our perceptual threshold. We mistook certainty for completion because articulation stabilized within a limited frame. We mistook progress for arrival because tension temporarily balanced.

But the structure was always dynamic.

This realization does not unsettle. It clarifies.

If nothing is absolutely still, then instability is not abnormal. If tension persists, then strain is not failure. If recalibration is constant, then revision is not weakness.

The field is alive because it is dynamic.

Uncertainty is not an intrusion into stability. It is the signature of motion within finite form.

After argument, what remains is simplicity.

Existence operates through tension held in balance. Alignment unfolds through delay. Resolution occurs within threshold.

And beneath every stabilization, motion continues.

The realization is not dramatic. It is quiet.

Nothing was ever fully certain. Nothing was ever fully still.

Yet everything persists — not by eliminating tension, but by holding it.

21.1 — Nothing Is Absolutely Still

Stillness is a perceptual convenience.

We look at a mountain and call it immovable. We look at a building and call it stable. We look at a person sitting quietly and call them at rest.

But beneath every appearance of stillness, motion persists.

At the physical level, matter is never static. Atoms vibrate within lattices. Particles fluctuate within fields. Energy redistributes continuously. What appears solid is dynamic equilibrium sustained within thresholds too subtle for ordinary perception.

Stillness is resolution below visibility.

At the biological level, a resting body is never inactive. Cells metabolize. Neurons fire. Hormones circulate. Micro-adjustments occur without conscious awareness. Stability is maintained through constant regulation.

At the psychological level, a calm mind is not empty of movement. Thoughts arise and dissolve. Memory reorganizes. Attention shifts microscopically. Emotional tone adjusts in response to internal and external cues.

At the civilizational level, even during apparent peace, negotiation persists. Economic exchanges continue. Institutional recalibrations occur. Cultural narratives evolve incrementally.

Nothing is absolutely still.

What we interpret as stability is motion organized within manageable amplitude.

This insight dissolves a common misunderstanding. We tend to equate stability with absence of tension. But absence of tension would mean absence of differentiation. And absence of differentiation would mean absence of structure.

Structure requires contrast. Contrast requires variation. Variation requires motion.

If a system were perfectly static — no internal redistribution, no fluctuation, no recalibration — it would not function. It would not adapt. It would not respond. It would not persist in a dynamic environment.

Persistence depends on controlled movement.

Earlier in this book, stability was described as balanced strain. That balance does not eliminate motion. It regulates it. The field never collapses into absolute stillness because collapse would terminate articulation itself.

Even in moments of profound calm — personal or civil — microscopic adjustment continues. Dynamic equilibrium is maintained through ongoing redistribution of distortion.

This realization reframes anxiety.

If nothing is absolutely still, then fluctuation is not abnormal. Change is not intrusion. Micro-instability does not signal failure.

It signals life.

The mountain erodes slowly. The institution reforms gradually. The mind shifts subtly.

Motion is continuous, even when imperceptible.

Absolute stillness would require zero distortion and zero delay. But distortion and delay are structural prerequisites for existence. They enable interaction. They allow alignment to occur over time.

Therefore, what we call stability is motion held within boundary.

And once this is understood, the fear of movement softens.

Nothing was ever frozen in perfection. Nothing was ever meant to be.

The field breathes through constant recalibration.

Stillness is not absence of motion. It is motion integrated gracefully.

21.2 — Stability Is Tension Held Gracefully

If nothing is absolutely still, then stability must be redefined.

Stability is not rigidity. It is not suppression of disturbance. It is not elimination of competing forces.

It is tension held without collapse.

A bridge stands not because force is absent, but because force is distributed. Load is absorbed, redirected, balanced. Remove tension entirely and the structure loses purpose. Overwhelm it and it fractures.

The same principle extends beyond engineering.

A stable mind does not lack conflicting thoughts. It can hold opposing perspectives without disintegration. It can experience fear without paralysis, doubt without surrender, desire without compulsion. The tension remains, but it is organized.

Grace enters here.

Grace is not the absence of strain. It is the ability to distribute strain intelligently.

A mature person does not eliminate uncertainty. They carry it without agitation. They make decisions knowing incompletion persists. They adjust when misalignment appears. They do not demand final resolution before acting.

At the civil level, stability does not mean unanimous agreement. It means managed disagreement. Competing interests coexist because institutional frameworks distribute tension across channels — legal, economic, cultural. When distribution fails, polarization intensifies.

Graceful stability requires elasticity.

Rigid systems shatter under strain because they resist redistribution. Flexible systems adapt because they accept motion within boundary. Elasticity does not deny tension; it integrates it.

This reframing alters expectation.

If stability required absence of strain, it would be unattainable. Strain is structural. Distortion arises naturally in dynamic fields. Competing deltas persist. What matters is not elimination, but organization.

A stable society debates. A stable relationship negotiates. A stable ecosystem fluctuates within range.

The common feature is not stillness — it is balanced adjustment.

Earlier in this book, uncertainty was described as pre-resolved distortion within finite thresholds. Stability does not remove that distortion. It contains it within tolerable amplitude.

Graceful tension means accepting that alignment is ongoing.

There is no final rest state. There is no permanent equilibrium immune to shift. What exists instead is continuous recalibration performed without panic.

When tension is denied, it accumulates invisibly. When tension is acknowledged, it can be distributed.

Stability, therefore, is not a static condition achieved once and preserved indefinitely. It is a practice — of adjustment, of listening, of recalibrating before fracture.

Grace lies in accepting motion without demanding immobility.

Nothing collapses simply because tension exists. Collapse occurs when tension cannot be reorganized.

To live within uncertainty is to develop this capacity — to hold competing forces without immediate reaction, to allow delay without anxiety, to adjust without drama.

Stability is not silence of disturbance. It is disturbance held intelligently.

And that holding — quiet, adaptive, continuous — is what allows dynamic systems to endure.

21.3 — Participation Without Control

If nothing is absolutely still, and stability is tension organized rather than eliminated, then the idea of control requires reconsideration.

Control suggests dominance over the field.

Total predictability. Total direction. Total authority over outcome.

But in a dynamic system governed by finite thresholds, such dominance is illusion.

We influence. We adjust. We participate.

We do not command the entire geometry.

Every action we take enters a network of interacting variables. Decisions ripple outward, encounter other forces, and return in altered form. The result is rarely identical to the intention. Not because effort fails, but because the field is larger than any single articulation.

Participation acknowledges this.

To participate is to act within constraint, aware that alignment is partial. It accepts that distortion will not vanish, that recalibration will follow, that consequence extends beyond immediate design.

This is not helplessness.

It is maturity.

Control seeks certainty before movement. Participation accepts uncertainty and moves responsibly within it.

In personal life, control manifests as the demand for predictable outcomes — from relationships, careers, reputation. When unpredictability appears, frustration arises. But participation recognizes that others are independent centers of alignment. Their internal deltas cannot be fully governed.

In civil life, control appears as rigid ideology — attempts to impose singular articulation across diverse systems. Participation, by contrast, negotiates. It adapts. It integrates feedback.

In science, control would require absolute prediction. Participation investigates within threshold, refines models, and accepts provisionality.

In technology, control promises optimization without residual strain. Participation designs tools while acknowledging limits and unintended consequences.

The illusion of control collapses when thresholds are crossed.

Unexpected events occur. Systems behave nonlinearly. External variables intervene. The field reasserts its complexity.

Participation does not eliminate responsibility. It deepens it.

When one accepts that complete control is impossible, attention shifts to quality of interaction. Instead of forcing the field, one aligns with it. Instead of demanding static outcomes, one monitors shifting equilibrium.

This aligns with everything established earlier: uncertainty is structural. Therefore, action must be adaptive.

Participation without control means:

- Acting with awareness of limitation.
- Accepting recalibration as normal.
- Avoiding rigidity in expectation.
- Holding responsibility without claiming omnipotence.

This position does not reduce agency. It clarifies it.

Agency exists locally. Influence exists relationally. Outcome exists collectively.

The field is dynamic. We are nodes within it.

To participate is to engage intelligently within motion, without demanding stillness.

Control dissolves under scrutiny. Participation remains.

And participation — deliberate, adaptive, aware — is how finite beings live within dynamic equilibrium without collapse.

21.4 — Living Within Dynamic Equilibrium

To live within dynamic equilibrium is not to withdraw from the world. It is to move within it without demanding permanence.

Dynamic equilibrium means tension persists, but it does not overwhelm. Motion continues, but it does not destabilize. Recalibration occurs, but it does not threaten identity.

It is the acceptance that balance is not fixed — it is maintained.

In earlier chapters, equilibrium was described as distributed strain within finite thresholds. That description now becomes personal.

Living within dynamic equilibrium means:

You expect fluctuation. You expect revision. You expect incomplete articulation.

And because you expect it, you are not shocked by it.

A decision does not eliminate doubt completely. A relationship does not eliminate conflict permanently. A society does not eliminate disagreement indefinitely.

Instead, each maintains coherence through ongoing adjustment.

Dynamic equilibrium is not passive. It requires attentiveness. It requires sensing when strain is accumulating beyond threshold and responding before fracture. It requires allowing motion instead of resisting it.

In practical terms, this becomes a posture.

You make decisions knowing they are provisional. You build systems knowing they will evolve. You hold convictions knowing refinement may follow.

This is not relativism. It is structural realism.

Certainty becomes functional rather than absolute. Stability becomes adaptive rather than rigid. Confidence becomes grounded rather than inflated.

Living this way reduces anxiety because it aligns expectation with structure. If uncertainty is fundamental, then surprise is not betrayal. It is continuation.

You do not demand stillness from a moving field. You synchronize with its rhythm.

Dynamic equilibrium also changes how one experiences time. Delay is no longer frustration alone; it is structural necessity. Resolution requires duration. Articulation unfolds through threshold. Patience becomes alignment rather than weakness.

The quiet realization is not dramatic. It does not require philosophical intensity. It is almost ordinary.

Nothing is absolutely fixed. Nothing is absolutely certain. Nothing is absolutely controllable.

Yet coherence persists.

The mind adapts. Systems recalibrate. Civilizations reorganize.

Living within dynamic equilibrium means recognizing that persistence does not require perfection — only redistribution of strain within tolerable range.

The field moves. We move with it.

Uncertainty remains, not as threat, but as constant companion of alignment.

And once that is understood, stability no longer means freezing the world.

It means moving within it — deliberately, responsibly, and without illusion of final rest.

Closing Reflection

Nothing in this book asked you to reject science. Nothing asked you to distrust progress. Nothing asked you to abandon precision.

It asked only for structural clarity.

We searched for certainty in knowledge, in technology, in institutions, in control. We expected that refinement would eventually silence unpredictability. We assumed that with enough data, enough computation, enough discipline, uncertainty would dissolve.

But it never did.

Not because we failed. Not because the models were flawed. Not because effort was insufficient.

It remained because it belongs to structure.

Nothing is absolutely still. Nothing is absolutely certain. Nothing is absolutely complete.

Yet everything persists.

Matter stabilizes through motion. Minds stabilize through tension. Societies stabilize through negotiation.

Equilibrium is not the absence of distortion. It is distortion organized within threshold.

Uncertainty was never an interruption in the system. It was the condition that made the system possible.

Without uncertainty, there would be no inquiry. Without delay, there would be no alignment. Without finitude, there would be no articulation.

The problem was never uncertainty itself.

The problem was believing it should disappear.

Once that belief softens, something else emerges — not resignation, not indifference, but composure.

You act without demanding permanence. You decide without requiring omniscience. You build without expecting finality.

The field remains dynamic.

And within that motion, stability is still possible — not as stillness, but as balanced participation.

Nothing was ever fully certain.

And that was never a flaw.

Epilogue — The World Was Never Certain

Before equations, before institutions, before instruments, the world was already in motion.

Stars formed and dissolved. Fields fluctuated. Matter reorganized. Energy redistributed.

Nothing waited for certainty to begin.

Long before humans measured time, processes unfolded within thresholds. Long before we named forces, interactions stabilized through dynamic balance. Long before we constructed models, the field operated through tension held within constraint.

The world was never certain.

It was structured. It was dynamic. It was finite.

When humans began to observe, uncertainty entered language. We called some events predictable and others unpredictable. We called some systems stable and others chaotic. We separated knowledge from ignorance and believed the boundary would shrink until certainty prevailed.

But the boundary did not shrink to zero.

Instead, it moved.

Each refinement exposed deeper structure. Each explanation revealed new variables. Each stabilization uncovered subtler fluctuation.

The world did not become more uncertain. We became more aware of its structure.

Even now, as civilization reaches unprecedented capability — global networks, artificial systems, instantaneous communication — the fundamental condition remains unchanged.

Alignment requires time. Resolution consumes bandwidth. Distortion redistributes.

No discovery has removed delay. No model has eliminated threshold. No technology has abolished finitude.

The world was never certain, not because it was disordered, but because it was alive with interaction.

Certainty, as we imagined it, would require static equilibrium — no competing deltas, no motion beneath perception, no emerging configuration. Such a world would not evolve. It would not differentiate. It would not produce inquiry.

It would not produce us.

Uncertainty is not a late complication of advanced civilization. It is the original condition of dynamic existence. From the smallest fluctuation to the largest structure, articulation has always occurred within constraint.

The mountain was never still. The ocean was never fixed. The mind was never absolute.

And yet coherence persists.

Systems stabilize locally. Organisms adapt. Civilizations endure.

Not because uncertainty disappears, but because it is integrated.

Perhaps the deepest realization is this:

We did not enter an uncertain world. We were born from one.

Our intelligence evolved within threshold. Our sciences matured within delay. Our societies formed through negotiation of competing forces.

The world was never certain — and that is precisely why it could unfold.

If certainty had been absolute, nothing would change. If thresholds did not exist, nothing would differentiate. If distortion did not arise, nothing would articulate.

Uncertainty was not added to reality. It is woven into it.

And once this is seen, the demand for final resolution softens.

The goal is no longer to eliminate uncertainty. It is to understand its structure.

The world continues as it always has — dynamic, finite, recalibrating.

Not perfectly resolved. Not absolutely predictable.

Yet enduring.

The world was never certain.

It never needed to be.

Appendices

Uncertainty is not ignorance.

Appendix A — Core Structural Definitions

The arguments in this book rely on a small set of structural terms.

These terms are not decorative. They are architectural. Each one describes a recurring pattern that appeared across physics, psychology, civilization, and technology. Without clarity in language, continuity across scales can appear abstract. With clarity, the structure becomes visible.

The definitions that follow are not metaphors. They are functional descriptions of how dynamic systems operate under finite conditions.

They are meant to be read slowly.

Not to persuade — but to stabilize understanding.

1. Structured Vacuum Energy (SVE)

Structured Vacuum Energy refers to the continuous foundational field within which all articulation occurs. It is not empty space, nor a collection of particles, but a dynamic substrate capable of sustaining distortion, redistribution, and alignment.

SVE is structured in the sense that it permits geometry, interaction, and threshold behavior. It is continuous in that no absolute void exists within the system.

All phenomena emerge as patterned distortion within SVE.

2. Distortion

Distortion is deviation from local equilibrium within the field.

It is not disorder. It is not malfunction.

Distortion is the structural difference that allows articulation to occur. Without distortion, no differentiation would exist. Without differentiation, no interaction would unfold.

Distortion is the origin of motion, change, and process.

3. Alignment

Alignment is the redistribution of distortion toward local equilibrium.

It is the structural response of the field to deviation. Alignment does not eliminate distortion absolutely; it reorganizes it within tolerable thresholds.

All stabilization events are alignment processes.

4. Alignment Threshold Time (ATT)

Alignment Threshold Time refers to the finite duration required for distortion to reorganize into temporary equilibrium.

No resolution is instantaneous. No alignment is infinite.

ATT defines the temporal bandwidth within which articulation occurs. It explains why delay is structural and why uncertainty persists during reorganization.

5. Finite Bandwidth

Finite bandwidth is the limited capacity of any system — physical, biological, cognitive, civil, or technological — to process and articulate competing distortions simultaneously.

Because bandwidth is finite, articulation is always partial. Resolution occurs locally, never exhaustively.

Finite bandwidth guarantees structured incompletion.

6. Dynamic Equilibrium

Dynamic equilibrium is stability achieved through continuous redistribution of distortion within finite thresholds.

It is not stillness. It is tension held in balance.

All persistent systems — from atoms to civilizations — operate through dynamic equilibrium rather than static perfection.

7. Residual Strain

Residual strain is distortion that remains unarticulated after local alignment.

No resolution exhausts the field completely. Residual strain persists and may re-emerge when thresholds shift.

Residual strain explains why stability is provisional and why recalibration is continuous.

8. Local Certainty

Local certainty is exhaustion of articulation within a defined bandwidth and frame.

It does not imply global completeness. It indicates that within current thresholds, distortion has stabilized sufficiently to permit coherent action.

All scientific and operational certainty is local.

9. Structured Incompletion

Structured incompletion is the condition in which systems remain partially unresolved due to finite bandwidth and dynamic interaction.

It is not ignorance. It is not error.

It is the inherent limitation of finite alignment operating within continuous structure.

Uncertainty emerges from structured incompletion.

10. Participation

Participation is action taken within dynamic equilibrium, acknowledging finite control and inevitable recalibration.

It replaces the illusion of total dominance with intelligent engagement under constraint.

Participation is the practical posture aligned with structural uncertainty.

Appendix B — Cross-Scale Structural Mapping

Uncertainty appears across multiple domains. Not as analogy, but as structural recurrence. This appendix displays that recurrence directly.

The same alignment geometry appears in:

- Physical systems
- Biological systems
- Cognitive systems
- Civilizational systems
- Technological systems

The terminology changes. The structure does not.

I. Core Structural Recurrence

Structural Principle across domains:

Distortion

- Physical: Field deviation
- Biological / Cognitive: Neural conflict, emotional tension
- Civil: Social disagreement
- Technological: System load, data imbalance

Alignment

- Physical: Energy redistribution

- Biological / Cognitive: Decision-making, regulation
- Civil: Policy formation, negotiation
- Technological: Output generation, optimization

Threshold

- Physical: Interaction limit
- Biological / Cognitive: Attention, processing capacity
- Civil: Institutional capacity
- Technological: Compute, memory constraints

Delay (ATT)

- Physical: Propagation time
- Biological / Cognitive: Cognitive processing time
- Civil: Legislative cycle
- Technological: Processing latency

Residual Strain

- Physical: Fluctuation, instability
- Biological / Cognitive: Doubt, unresolved tension
- Civil: Structural vulnerability
- Technological: Drift, model instability

Dynamic Equilibrium

- Physical: Stable oscillation
- Biological / Cognitive: Psychological balance
- Civil: Political stability
- Technological: System stability

This mapping is not metaphorical. Each domain exhibits:

- Finite processing capacity
- Competing distortions
- Time-bound alignment
- Incomplete stabilization

Uncertainty emerges from the same geometry.

II. Scale-Invariant Pattern

Across all domains, the following sequence repeats:

1. Distortion emerges
2. Alignment attempts redistribution
3. Threshold limits articulation
4. Residual strain persists
5. Recalibration becomes necessary

This pattern is observable in:

- Quantum measurement
- Emotional conflict
- Ideological polarization
- Machine learning convergence
- Ecosystem adaptation

The recurrence indicates structural continuity, not domain-specific anomaly.

III. Certainty Across Scales

Certainty means different things depending on the domain:

- **Physics:** Predictive stability within a model, because the model operates under finite conditions.

- **Mind:** Decision conviction, bounded by cognitive bandwidth.
- **Society:** Institutional consensus, shaped by shifting social thresholds.
- **Technology:** Deterministic output, constrained by architecture.

Certainty represents stabilized articulation within defined bandwidth. It never implies global elimination of distortion.

IV. Failure Modes Under Misinterpretation

When structural uncertainty is misunderstood, the following failures occur:

- Misinterpretation: Uncertainty = ignorance → Consequence: Endless search for total information
- Misinterpretation: Uncertainty = randomness → Consequence: Abandonment of structural analysis
- Misinterpretation: Uncertainty = error → Consequence: Overcorrection and rigidity
- Misinterpretation: Stability = stillness → Consequence: System brittleness
- Misinterpretation: Control = dominance → Consequence: Collapse under complexity

These failures occur because structure is denied.

V. Unified Structural Insight

Across scales:

- Nothing operates under infinite bandwidth.
- Nothing resolves instantaneously.

- Nothing eliminates distortion completely.

Dynamic systems persist through distributed tension. Uncertainty is not domain-specific. It is scale-invariant.

VI. Why This Matters

This mapping is not philosophical decoration. It implies:

- Scientific humility without relativism
- Institutional design that expects recalibration
- Technological development aware of drift
- Psychological maturity that accepts tension

The same structural geometry governs all dynamic systems. Understanding this continuity prevents overreach. It stabilizes expectation. It reframes uncertainty from anomaly to architecture.

Appendix C — Common Misinterpretations

When a structural reframing is proposed, it is often interpreted through existing conceptual habits.

This appendix addresses common misunderstandings that may arise from reading this work. The clarifications below do not extend the theory. They refine its boundaries.

I. Does This Deny Determinism?

No.

Determinism concerns whether events follow lawful structure. This framework does not reject lawful structure. It affirms it.

What is challenged is the assumption that lawful structure implies infinite articulability within finite systems.

Even in deterministic systems, finite bandwidth and temporal thresholds limit operational articulation. A system may be governed by coherent laws and still exhibit uncertainty under finite resolution.

Determinism in principle does not eliminate threshold in practice.

Uncertainty here does not imply absence of law. It implies finite alignment under law.

II. Does This Claim Everything Is Relative?

No.

Relativism suggests absence of structural grounding. This framework proposes the opposite: structural continuity across domains.

Local certainty exists. Models can be reliable. Decisions can be justified. Institutions can function coherently.

What is rejected is absolute exhaustion of distortion.

Certainty is contextual, not arbitrary. Bounded, not groundless.

Acknowledging threshold does not dissolve truth. It situates it.

III. Is Uncertainty Being Romanticized?

No.

Uncertainty is not celebrated as virtue. It is described as condition.

It generates discomfort. It produces tension. It requires recalibration.

But discomfort does not make it defective.

Romanticizing uncertainty would imply valuing instability for its own sake. This framework values structured equilibrium. It recognizes that equilibrium necessarily includes residual strain.

Uncertainty is not glorified. It is normalized.

IV. Is This Anti-Science?

No.

Science operates by refining articulation within finite bandwidth. The recognition of structural incompletion strengthens scientific discipline.

Scientific certainty is local and operational — and that is sufficient for progress. Expecting absolute finality from science misrepresents its method.

This framework supports:

- Iterative refinement

- Model revision
- Threshold awareness
- Intellectual humility

It does not undermine empirical rigor. It contextualizes it.

V. Is This Anti-Technology?

No.

Technology increases articulation bandwidth. It accelerates alignment. It reduces ignorance. It extends capability.

What it does not do is eliminate structural finitude.

Understanding this prevents overextension of technological promises. It encourages design that anticipates recalibration, drift, and unintended consequence.

Technology remains powerful — but not absolute.

VI. Does This Undermine Human Agency?

No.

Agency operates locally within dynamic systems. It does not require infinite control to remain meaningful.

In fact, recognizing structural uncertainty deepens responsibility. When outcomes cannot be guaranteed, ethical awareness increases.

Participation replaces illusion of dominance.

Agency persists — within threshold.

VII. Does This Imply Collapse Is Inevitable?

No.

Dynamic equilibrium does not imply fragility. It implies continuous recalibration.

Systems collapse when thresholds are denied, strain accumulates invisibly, and redistribution becomes impossible.

Understanding structural uncertainty enhances resilience.

Collapse is not destiny. It is mismanaged distortion.

VIII. Is This a Metaphysical Claim?

No metaphysical assertion is required.

The framework describes observable structural recurrence:

- Finite processing
- Time-bound alignment
- Persistent residual strain

These are empirically identifiable conditions across domains.

The conclusions arise from structural continuity, not speculative metaphysics.

IX. Is Absolute Certainty Impossible in Principle?

Absolute certainty would require infinite articulation capacity.

This framework observes that all dynamic systems operate under finite thresholds.

If one imagines an infinite system beyond all constraint, absolute certainty may be conceivable abstractly. But such a system would not resemble dynamic existence.

The claim concerns structured, finite systems.

X. What This Book Does Not Claim

It does not claim:

- That knowledge is futile
- That prediction is meaningless
- That stability is unattainable

- That progress is illusion
- That systems cannot improve

It claims only:

Improvement operates within structure. Structure includes finitude. Finitude guarantees structured incompletion.

Uncertainty follows from that.

Closing Clarification

This work does not seek to unsettle confidence. It seeks to align expectation with structure.

Uncertainty is not introduced by this theory. It was already present. This framework simply makes its geometry visible.

Appendix D — Operating Within Structural Uncertainty

If uncertainty is structural, then the question is no longer whether it can be eliminated.

The relevant question becomes:

How should systems — scientific, institutional, technological, and personal — operate under this condition?

This appendix does not offer instruction. It offers orientation.

I. Scientific Posture — Precision Without Absolutism

Scientific inquiry refines articulation within finite thresholds.

Operating under structural uncertainty requires:

- Recognizing model scope
- Distinguishing operational certainty from global completeness
- Designing experiments that anticipate boundary conditions
- Remaining open to recalibration without destabilizing prior progress

Scientific maturity is not weakened by incompletion. It is strengthened by awareness of it.

Certainty becomes functional, not final.

Revision becomes structural, not embarrassing.

II. Institutional Design — Expecting Recalibration

Institutions often fail when they assume permanence.

Structural uncertainty implies:

- Thresholds shift
- Capacity saturates
- Strain redistributes
- Conditions evolve

Institutions designed with flexibility — feedback channels, adaptive review, distributed load — persist longer.

Rigidity accumulates residual strain.

Dynamic governance does not signal instability. It signals alignment within motion.

III. Technological Development — Designing for Drift

Technological systems operate within finite architecture.

Operating under structural uncertainty means:

- Expecting model drift
- Monitoring threshold saturation
- Anticipating unintended interactions
- Avoiding claims of final optimization

No system reaches absolute resolution. Scaling increases capability, but not infinite articulation.

Resilience becomes more valuable than perfection.

Systems should be built for recalibration.

IV. Leadership and Decision-Making — Acting Without Omniscience

Leadership often collapses under the illusion of total control.

Operating within structural uncertainty requires:

- Acting with incomplete information
- Accepting residual doubt
- Distributing decision strain
- Allowing adaptive revision

Decisiveness does not require absolute certainty.

It requires sufficient local stability to proceed responsibly.

The demand for total clarity before action creates paralysis. The assumption of total clarity creates overreach.

Structural awareness stabilizes both.

V. Personal Orientation — Tension Without Panic

At the individual level, uncertainty manifests as:

- Doubt
- Delay
- Internal conflict
- Fear of unpredictability

Operating structurally does not remove these experiences.

It reframes them.

Conflict signals competing internal deltas. Delay signals alignment in process. Doubt signals incomplete articulation.

None of these indicate defect.

Personal maturity involves:

- Holding tension without immediate reaction

- Revising without ego collapse
- Acting without final guarantee
- Accepting recalibration as normal

Stability is not the absence of fluctuation. It is regulated fluctuation.

VI. Progress — Advancement Without Illusion

Progress expands bandwidth. It does not eliminate finitude.

Operating structurally means:

- Expecting new vulnerabilities with new capabilities
- Recognizing that every solution redistributes distortion
- Designing advancement with adaptive capacity

Technological acceleration increases complexity. Complexity increases threshold sensitivity.

Sustainable progress integrates recalibration into its architecture.

VII. The Structural Discipline

Across all domains, operating within structural uncertainty involves a shared discipline:

- Do not equate uncertainty with failure.
- Do not equate stability with stillness.
- Do not equate control with dominance.
- Do not equate certainty with completeness.

Instead:

- Treat articulation as local.
- Treat equilibrium as dynamic.

- Treat recalibration as expected.
- Treat incompletion as structural.

This discipline reduces overreaction.
It increases durability.

VIII. The Quiet Alignment

Nothing in this appendix demands transformation.

It suggests a shift in expectation.

Systems persist not because they eliminate uncertainty, but because they organize it.

Operating within structural uncertainty is not heroic.

It is realistic.

It allows science to refine, institutions to adapt, technology to mature, and individuals to act — without demanding finality.

The field remains dynamic.

Operating within it requires composure, not conquest.

Author Positioning Note

This work does not emerge from a single discipline.

It is written at the intersection of physics, systems thinking, cognitive analysis, institutional structure, and technological design. The intent is not to compete with established domains, but to identify a recurring structural pattern visible across them.

The framework presented here — articulated through concepts such as distortion, alignment, finite bandwidth, and dynamic equilibrium — is offered as a cross-domain continuity model. It does not replace existing scientific theories, nor does it seek to invalidate operational methodologies. It seeks to clarify a shared structural condition underlying them: finite alignment within dynamic systems.

The author approaches uncertainty not as an anomaly within physics, psychology, or civilization, but as a foundational property of structured existence. The argument is architectural rather than ideological.

This book is positioned as a foundational reference within a broader research trajectory that examines scale-invariant structural principles across domains. It invites dialogue, refinement, and critical engagement — not agreement by assertion.

The intention is clarity.

Uncertainty is not treated as a philosophical abstraction, nor as a statistical inconvenience, but as structural continuity.

The framework remains open to refinement, extension, and formal critique.

It is offered not as conclusion, but as orientation.

Acknowledgements

This book was not written in isolation, even though much of its thinking unfolded in solitude.

I am grateful to the scientific tradition that trained the mind to question without fear and to refine without attachment. Generations of researchers, physicists, philosophers, engineers, and systems thinkers have built the intellectual environment within which structural inquiry is possible. Whether in agreement or disagreement, their rigor shaped the discipline behind these pages.

My students deserve quiet acknowledgment. Teaching forces clarity. When an idea cannot be explained simply, it is not yet understood. Many refinements in this manuscript emerged from the necessity of explaining structure without excess language.

I also acknowledge readers across previous works who engaged critically, questioned assumptions, and pushed for sharper articulation. Their challenges strengthened the framework.

To my family, who tolerate long silences and unfinished conversations while ideas stabilize, thank you. Stability in life makes exploration possible.

Finally, I acknowledge uncertainty itself — not as an adversary, but as the condition that made inquiry necessary.

This work is part of an ongoing exploration. Any clarity it offers rests on countless visible and invisible influences that made it possible.

About the Author

Sandeep J. Chavan is an independent researcher, educator, and interdisciplinary thinker working at the intersection of physics, systems theory, and human cognition. Trained in engineering and deeply engaged with foundational questions in science and philosophy, his work focuses on identifying structural continuity across physical, psychological, and civil domains.

He is the architect of a broader research trajectory that examines finite alignment, dynamic equilibrium, and scale-invariant structural principles within complex systems. Rather than confining inquiry to a single discipline, his approach integrates scientific reasoning with systems analysis to explore how uncertainty, stability, and articulation operate across domains.

As a teacher, he emphasizes conceptual clarity over formulaic repetition, encouraging structural understanding rather than memorization. His writing reflects this orientation — analytical, cross-disciplinary, and grounded in first principles.

This book forms part of a continuing body of work dedicated to reframing foundational assumptions about uncertainty, certainty, and dynamic existence.

Uncertainty — In a Nutshell

Through Structural Quotes and Laws

Core Structural Statements

- Nothing is absolutely still.
- Stability is tension held in balance.
- Certainty is local exhaustion of articulation.
- Resolution is finite.
- Delay is structural.
- Incompletion is inherent.
- Control is limited.
- Participation is necessary.

Chavanian Laws of Uncertainty

Law 1 — The Law of Finite Alignment

Every dynamic system operates under finite articulation bandwidth.

Absolute resolution is impossible within finite thresholds.

Law 2 — The Law of Structural Delay

All alignment requires duration.

No resolution occurs without temporal threshold.

Law 3 — The Law of Residual Strain

Every stabilization leaves unarticulated distortion.

Residual strain guarantees future recalibration.

Law 4 — The Law of Dynamic Equilibrium

Stability is not absence of tension.

It is balanced redistribution of tension within tolerable range.

Law 5 — The Law of Local Certainty

Certainty exists only within defined scope and bandwidth.

Global completeness is unattainable in dynamic systems.

Law 6 — The Law of Scale Continuity

The same structural geometry governs physical, cognitive, civil, and technological systems.

Uncertainty is scale-invariant.

Law 7 — The Law of Threshold Sensitivity

As complexity increases, threshold sensitivity increases.

Expansion of capability increases exposure to instability.

Law 8 — The Law of Non-Absolute Control

Influence is local.

Total dominance over dynamic systems is structurally impossible.

Condensed Structural Insight

- Uncertainty is not ignorance.
- Uncertainty is not randomness.
- Uncertainty is not error.
- Uncertainty is structured incompletion under finite alignment.

The Final Compression

Uncertainty is the boundary condition of finite systems participating in dynamic equilibrium.

It is not the problem.

Misunderstanding it is.

Disclaimer

This book presents a structural framework for understanding uncertainty across physical, cognitive, civil, and technological domains. The concepts discussed are interpretative and analytical in nature and are offered for intellectual exploration.

Nothing in this work should be interpreted as medical, legal, financial, engineering, or policy advice. Readers are encouraged to consult qualified professionals before applying any ideas discussed herein to practical decision-making within specialized domains.

The scientific discussions presented aim to provide structural interpretation and cross-domain continuity. They do not claim to replace established empirical models, peer-reviewed research, or domain-specific methodologies. Where scientific concepts are referenced, they are used for structural analysis rather than formal technical derivation.

This book does not advocate rejection of existing scientific consensus, institutional frameworks, or technological systems. It proposes an architectural reframing of uncertainty within dynamic systems.

All interpretations remain open to scholarly dialogue, refinement, and critique.

Statement of Originality

This work represents an original conceptual synthesis developed by the author.

While it draws upon established scientific principles, systems theory, and interdisciplinary knowledge, the structural framing presented — including the articulation of finite alignment, dynamic equilibrium, structured incompletion, and scale-invariant uncertainty — reflects the author's independent research and analytical development.

All interpretations, structural mappings, and cross-domain integrations are the result of original thought and synthesis. Any references to established theories or scientific principles are acknowledged as part of the broader intellectual landscape within which this work participates.

This manuscript does not knowingly reproduce proprietary frameworks, copyrighted materials, or unpublished research without attribution. The structural terminology and continuity model presented herein form part of the author's broader research trajectory and intellectual body of work.

This book is offered as an independent contribution to ongoing interdisciplinary dialogue on uncertainty and dynamic systems.

Further Reading

From the Chavanian Ecosystem

The themes explored in *Uncertainty* extend across multiple works within the broader Chavanian research framework. Readers interested in deepening structural, philosophical, and applied dimensions may refer to the following:

Foundations of Universal Energy Dynamics (UED)

Black Hole Don't Exist, A structural re-examination of gravitational collapse, energy redistribution, and alignment thresholds within cosmic systems.

Chavan–Heisenberg Ripple Uncertainty Principle, An extension of quantum indeterminacy through ripple geometry and finite articulation bandwidth.

Chaos: A Gentle Guide to the Unresolvable, A structural exploration of incompletion, instability, and dynamic balance across domains.

A Brief History of Intelligence, Tracing the evolution of articulation, awareness, and structured cognition within dynamic systems.

Mind, Consciousness, and Alignment

The Active Mind, An examination of cognitive tension, internal deltas, and psychological equilibrium.

Energy in Action: The Soul Demystified, A structural account of consciousness as organized energy within finite thresholds.

Subconsciousness: The Chosen Disconnection, Exploring layered awareness, delay, and structured incompletion within the human mind.

The Mind That Won't Stop, An analysis of persistent cognitive motion and threshold saturation.

Human Systems and Civilization

The Human Game, A study of competition, cooperation, and structural uncertainty within social architecture.

Judge Outside the Courtroom, An inquiry into justice, procedural thresholds, and epistemic limitation.

Aligned Father, Misaligned Dad, Exploring generational uncertainty, expectation, and structural maturity.

Technology and Artificial Systems

Why Do AI Systems Hallucinate?, A structural diagnosis of coherence, completion pressure, and finite model articulation.

The Human Machine, Examining biological and technological convergence through field-based structural continuity.

The Real Story of AI, A reframing of artificial intelligence beyond anthropomorphic interpretation.

Philosophical and Existential Continuity

Aham Brahmasmi, An inquiry into identity, consciousness, and structural participation.

Bubbles: Become a Drop, A metaphorical articulation of individuality within continuous field structure.

Breath, A reflection on rhythm, delay, and dynamic equilibrium in lived experience.

Core Structural Thread

Across these works, recurring themes include:

- Finite alignment
- Dynamic equilibrium
- Residual strain
- Structural delay
- Scale continuity
- Participation over dominance

Uncertainty functions as a foundational articulation of these principles.

Readers may approach the ecosystem in any order; the structural geometry remains continuous.

Don't miss out!

Visit the website below and you can sign up to receive emails whenever Sandeep Chavan publishes a new book. There's no charge and no obligation.

https://books2read.com/r/B-A-EPGPC-OVOBJ

Connecting independent readers to independent writers.

Did you love *Uncertainty – The Pre-Resolved Strain*? Then you should read *Why Do AI Systems Hallucinate?*[1] by Sandeep Chavan!

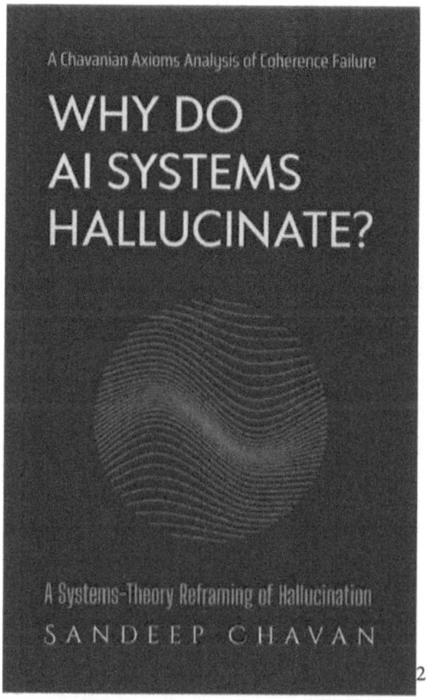

[2]

Why do AI systems hallucinate—even when trained on massive datasets and refined with advanced alignment techniques?

Most explanations blame insufficient data, model scale, overfitting, or probabilistic uncertainty. But what if hallucination is not an implementation flaw at all? What if it is a structural outcome of how generative systems define stability?

In *Why Do AI Systems Hallucinate?*, Sandeep Chavan presents a systems-theory reframing of hallucination grounded in coherence dynamics rather than surface performance metrics. Using the Chavanian Axioms as a diagnostic lens, this book argues that hallucination emerges from a deeper architectural commitment:

1. https://books2read.com/u/479Jvg

2. https://books2read.com/u/479Jvg

forced resolution under continuity constraints without true dissipation.

This book does not propose a new architecture. It does not offer quick technical patches. Instead, it asks a more foundational question: under what structural conditions does hallucination inevitably converge?

Through accessible but rigorous analysis, the book explores:

Why hallucination is misunderstood as "wrong information"The difference between fluency and truthHow scaling amplifies residue rather than eliminating itWhy fine-tuning and RLHF reshape behavior but not equilibriumWhy probabilistic confidence cannot detect structural incoherenceWhat a non-hallucinating system would require—axiomatically

Rather than treating hallucination as a temporary engineering problem, this work positions it as a coherence outcome. When systems must always resolve, must always continue, and cannot suspend under uncertainty, fabrication becomes the most stable option.

For AI researchers, engineers, policymakers, and critical thinkers, this book offers a structural framework to test systems beyond accuracy benchmarks. It introduces coherence audits, residue diagnostics, and equilibrium stress tests—without prescribing specific code or architectures.

This is not an anti-AI book. It is a design-maturity book.

If AI is to be trusted in education, governance, healthcare, and decision-making, stability must precede performance. Optimization builds capability. Coherence builds trust.

This book challenges the field to shift from asking *how to reduce hallucination* to asking *why hallucination converges*.

Because the geometry of a system determines its failure mode.

Read more at https://www.sandeepjchavan.com/.

www.ingramcontent.com/pod-product-compliance
Lightning Source LLC
LaVergne TN
LVHW090259210926
848480LV00001B/74

* 9 7 9 8 2 3 3 5 4 0 3 4 9 *